Antique Phonograph
Accessories & Contraptions

Timothy C. Fabrizio and George F. Paul

Schiffer Publishing Ltd®
4880 Lower Valley Road, Atglen, PA 19310 USA

Frontispiece:
"Tell him that I love him, too and that my love is full and true."
Courtesy of John T. Hoffman, The Phonophile.

Designed by Bonnie M. Hensley
Cover design by Bruce M. Waters
Type set in Goudy Handtooled/Aldine721 BT

ISBN: 0-7643-1763-6
Printed in China
1 2 3 4

Published by Schiffer Publishing Ltd.
4880 Lower Valley Road
Atglen, PA 19310
Phone: (610) 593-1777; Fax: (610) 593-2002
E-mail: Info@schifferbooks.com
Please visit our web site catalog at **www.schifferbooks.com**
We are always looking for people to write books on new and related subjects. If you have an idea for a book please contact us at the above address.

This book may be purchased from the publisher.
Include $3.95 for shipping.
Please try your bookstore first.
You may write for a free catalog.

In Europe, Schiffer books are distributed by
Bushwood Books
6 Marksbury Ave.
Kew Gardens
Surrey TW9 4JF England
Phone: 44 (0) 20 8392-8585; Fax: 44 (0) 20 8392-9876
E-mail: Bushwd@aol.com
Free postage in the U.K., Europe; air mail at cost.

Dedication

To all those who have so graciously
assisted us in our ever-expanding endeavor.

Noset *The* Autostartstop

Contents

Introduction

The history of technology is a web of interconnected developments. One invention not only leads to another, as James Burke explored in *Connections* (Little, Brown & Company, 1978), but inspires a host of adjunct technology – accessories, appendages, "add-ons," "peripherals." Hence, the invention of the talking machine was a "big bang" which sent out streamers of inspiration into many related areas. Take the examples of cylinder or disc records.

Records, objects that store a *record* of sound vibrations for future reproduction, did not exist prior to Edison's development of the Phonograph in 1877. Illustration 1-1 shows a scrap of the earliest type of recording medium, a sheet of tinfoil. After a sheet of foil was removed from a Phonograph mandrel, it was not practical to reattach it, so any understanding of "record wear" did not exist until the introduction of replayable recordings, circa 1888. In fact, there was considerable effort expended to *increase* cylinder record wear around the turn-of-the-twentieth-century.

In an acoustic playback system, loudness depends to a great extent on how much pressure is exerted by the reproducing stylus on the surface of the record. During the 1890s, cylinder record volume was faint, and many people relied on ear-tubes to improve the listening experience. This was due in part to the lightweight construction of Phonograph and Graphophone reproducers. Therefore, supplemental weight to increase pressure on the record was an option subsequently appended to many factory reproducers. This, of course, decreased the usable life of the relatively soft brown wax cylinders then in use – but the phonograph record was yet such a novelty that most customers were probably unconcerned.

After 1901, however, the quality of both cylinder records and reproducers increased greatly. Cylinders were now mass-produced by molding, and were both harder in consistency and better recorded. In the United States, the Lambert Company introduced celluloid records, which were loud and incredibly durable. Cylinder phonograph reproducers now incorporated more weight by design, and ear-tubes became obsolete. The promotion of supplemental weights became occasional, such as the British attachment shown in illustration 2-67.

Disc records followed a somewhat different course. With the commercial blossoming of Berliner's Gramophone in the late 1890s, the intent of the manufacturer was to exert as much force as possible on the disc, in order to coax the most volume out of it. This was a big selling point of the Gramophone versus its cylinder competitors: the Gramophone was loud. Of course, the friction of the steel needle on the shellac surface of the record, the considerable weight of the soundbox (as a disc talking machine's reproducer is called) and inertia caused by the mass of the horn and traveling arm all contributed to record wear.

Even after most disc talking machine manufacturers adopted an "improved" design configuration that removed the weight of the horn from the record (the so-called "back-mounted" arrangement), record wear remained considerable. Consequently, gizmos that promoted prolonged disc life appeared, such as the "Velvet Tone" Needle Balance (illustration 3-68). There was even a preparation sold to decrease wear in *both* discs and cylinders, by lubricating their surfaces: "Record Food" (see illustration 2-83).

The talking machine record provoked technologies that both increased and decreased its physical wear. However, these were only two avenues of development that arose from the sound recording. There was an amazing array of horns available to amplify the acoustic record, as explored in Chapter One, and a multitude of storage choices, as explored in Chapter Four. Each element of the talking machine, from its spring motor to its polished cabinet, inspired bursts of inventive genius as brilliant and colorful as the nighttime sky on the Fourth of July.

Values in This Book

Preparing a value system for a large number of diverse items is a task fraught with concerns. Out of a need to simplify this system and make it as easy to use as possible, we will designate prices *only* for the "peripheral" items, which are the specific focus of this book. Though some talking machines will be pictured, they will not receive values. That is, where an accessory appears mounted on an instrument, the value refers to the accessory only. Items which are extremely uncommon will be rated "very rare" (VR).

Values will be expressed *in the individual captions* by the following letter codes:

A $1000. and up
B $500. to $999.
C $250. to $499.
D $150. to $249.
E $100. to $149.
F $50. to $99.
G $25. to $49.
H $5. to $24.

This might seem to be a long list of prices. However, you will find that most of the values will actually fall within categories D, E, F and G. Many illustrations contain a selection of objects. If these items are similar enough, a general value code will be shown. If there is sufficient disparity in the prices, they will be individually valued.

By employing price categories instead of specific dollar amounts, we leave ourselves open to various "slings and arrows" of discontent. It is our feeling that ours is the only effective approach to the subject. However, this does add an element of discretion to the use of our book.

In using the value codes, we advise *conservatism*. If you are trying to establish the value of an object in your possession, the natural instinct is to seize upon the highest price in the given range, regardless of con-

dition. This can be a mistake. The condition of the items in the book is nearly always excellent, but the condition of *your* item is the important issue. Condition can affect the value by up to 50% - and occasionally more. The bottom end of a price range has as much validity as the top, though it seems a less attractive prospect. Different examples of the same type of object may represent the entire range of a certain price category, due to condition. The ranges are also intended to be *comparative*. Certain objects in the "C" category, for instance, may *never* reach $499 no matter how perfectly preserved they are.

Beyond condition, many, many factors affect the price of an "antique." Exact worth is something that can not be fixed precisely in time. We can only hint at it. Our guide is intended to help you understand the value of the objects illustrated, but we vigorously discourage the use of the phrase, "It *books* for such-and-such a price." True worth is what the market will bear. Quoting books, auction results or what a shop is asking will probably not elevate the price higher than the general market will currently support. Furthermore, the antiques market is a "living" entity – it expands and contracts. On-line auction services have changed the environment and, in the manner of on-line stock trading, accelerated the extremes. You may notice that certain prices come *down*. This is indicative only of a cycle. In the long term, antiques are among the best investments.

One: Horns and Horn Accessories

The horn is the first thing most people notice about a talking machine – especially newcomers. Nationally syndicated antique "experts" have written tips for identifying Edison Phonographs by the type of horn. This, of course, is rubbish, but has often plagued general antique price guides. It is true that certain manufacturers (especially disc machine firms) offered little or no choice in horn equipment, but for the most part, the "horn-as-identifier" approach is about as accurate as attempting to date a CD player by the speakers attached to it.

Talking machine horns were one of the most widely manufactured aftermarket items in the industry. Rare was the dealer (especially of cylinder phonographs) who did not stock a selection of horns by diverse manufacturers. The 14" long horns that accompanied Edison Phonographs until October 1907 were very often discarded in favor of larger, more colorful examples – which, by the way, vastly improved the playback of recordings. The small horn usually supplied with larger Columbia cylinder Graphophones was frankly referred to in catalogues as a "recording horn." In other words, Columbia recognized the advantage of larger horns, and offered a variety of styles built by outside manufacturers.

Real choice in disc talking machine horn equipment did not come until around 1906 when both Victor and Columbia began offering horns of various materials, styles, and sizes (wood, brass, aluminum, steel, "straight," "flower," etc). Accessory horns had been available since the late

1-1.This yellow European papier mâché horn measures 12 3/4" x 21 3/4" long and is suggestive of *fin de siècle* Paris. It could have popped out of a poster by Lautrec. The Type "N" Graphophone to which it is mounted lacks the usual decal on the cabinet, a characteristic seen on several very early examples. *Courtesy of the Julien Anton collection.* (Value code: B)

1890s from outside firms, but the original-equipment disc machine horns were generally larger than those of cylinder machines, and produced inherently louder volume. Consequently, aftermarket horns for disc machines are far less common than cylinder phonograph versions.

The history of talking machine horns was detailed in one of our earlier books, *Antique Phonograph Gadgets, Gizmos, and Gimmicks* (Schiffer, 1999), and we will not replicate that work. However, before proceeding to the images in this chapter, a few general points should be made. Conscious of certain exceptions to these guidelines, we submit the following basic outline for dating a talking machine horn (though not always the phonograph to which it is attached!).

Horns of the 1890s were generally "tin" (actually tinned steel) of crude design, often omitting the bell or employing straight rather than curved bells. Many of the earliest horn bodies were double-seamed, that is, the body was made in two halves. Decoration throughout the decade was usually black enamel ("Japanning") with simple gilt stripes. Brass horns of the "funnel" style were available at increased cost, as were nickel-plated brass examples. By 1898, the bells of many horns were being made of curved metal. "Seamless" brass horns became available, resembling band instruments in appearance. These horns were expensive to manufacture and survived only until around 1903. Beginning in the 1890s, one of the premier manufacturers of horns was the Philadelphia firm of Hawthorne & Sheble.

Horns of the 1900-1904 period were primarily black painted, steel bodied with a seamed brass bell (or occasionally all brass). Glass horns appeared briefly, but predictable attrition has robbed us of most of them. Gilt stripes survived on the smallest, least expensive horns of this period (such as fit a "Q" Graphophone or "Gem" Phonograph). The Standard Metal Manufacturing Company became a prominent supplier of horns during this time, and even when unmarked, its products can be identified by a diamond-shaped hanger attachment. Newark, New Jersey was home to Standard, as well as the Tea Tray Company, which offered various sizes of trumpet-style, black bodied/brass belled horns adorned with an eye-catching red decal (see illustration 1-28). Interestingly, this decal announced "Accepted *Standard* Amplifying Horn For Talking Machines." The Standard Metal Manufacturing Company must certainly have resented the imitation of its name.

In 1904, the "Flower" or "Morning Glory" horn appeared, so-called because its paneled construction reflected a floral shape. Manufacturers began painting the horns in the colors of actual and imagined flowers, including the delicate blending of colors inside the "blossom." It wasn't long before curiously redundant painted images of flowers were available on the interior surfaces of "Flower" horns. The public responded enthusiastically, and horn manufacturers supplied a large variety of floral designs to suit any taste. So popular did the "Flower" horn become, that existing stocks of older, trumpet-style horns quickly became unwanted. Throughout 1905 and 1906, the Standard Metal Manufacturing Company offered specially decorated straight horns – with painted flowers on their black japanned bodies. We hypothesize that such embellishments were meant to clear out inventory, but in any event, these unusual hybrids had a short life. We believe that Standard was the only major firm to decorate straight horns in this way.

By 1907, cylinder machine horns made of fiber and wood had entered the marketplace. Aluminum was employed by Columbia for its small cylinder horns and some larger disc machine horns, but was a material far more commonly used in Europe. Metal horns with wood-

1-2. This otherwise typical 26" long tin horn of the 1890s was embellished with what appears to be a crude fleur-de-lis. Decorations other than stripes are unusual on horns of this type. *Courtesy of the Scott and Denise Corbett collection*. (Value code: C)

grained finish had a vogue before being overtaken in popularity by real wood, although painted grain continued to embellish many Hawthorne & Sheble and Edison paneled horns. Eccentric designs such as the Searchlight horn, Standard's "Lily" horn, and the short-lived "Tulip" horn appeared until 1908, when the combination of an economic recession and the introduction of larger factory-supplied horns by Edison and Columbia caused a downturn in the accessory cylinder horn market.

The last major talking machine horn development was the 1909 introduction of the Edison "Cygnet" ("swan") horn, with its characteristic gooseneck shape. An excellent acoustic design, it was usually offered in black enamel. Metal wood-grained examples were available, as were (after 1910) wooden-belled versions in oak, mahogany, and spruce built by the Sheip & Vandegrift ("Music Master") of Philadelphia.

The talking machine horn's salad days were the first decade of the twentieth century. As an acoustic device, it remained unequaled in sound quality by the inferior internal horns of Victrolas or myriad upright look-alikes until the introduction of exponential internal horns in 1925. The external horn fell from public favor due to aesthetics, not performance. Horns became "old fashioned," and "out of date" – the same attributes that endear them to thousands of collectors today. They are often beautifully proportioned, whimsically decorated, and masterfully crafted – yet talking machine horns represented a mere eyeblink in the history of sound reproduction.

Horn dimensions will usually be expressed with the width of the bell first, followed by the length of the horn.

1-3. A lesson in humility: when one of the authors first saw this brass horn for sale in 1980, mounted on a Columbia machine, he chuckled at the sight of a band instrument substituted for a genuine talking machine horn. "What ignorance!" he thought. However, a closer look revealed the engraving as seen in the next illustration, and what had been ridicule became revelation. Announced in the August 1899 issue of the trade magazine *Phonoscope*, this optional horn for the Berliner Gramophone was patented on April 4, 1899. The manufacturer, A. Leforestier, Son & Company, claimed that the horn, "Makes the volume of band and instrumental records as loud and natural as the original music." Measures 9 1/2" diameter x 22" net length. *Courtesy of Carol A. Veome.* (Value code: VR)

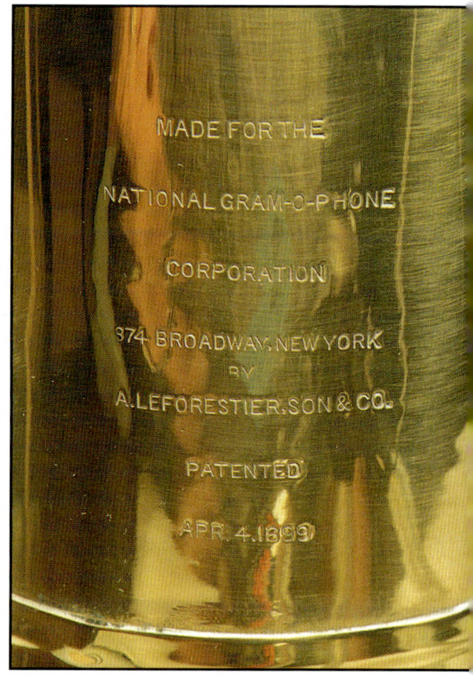

1-4. The engraving on the horn in previous illustration. It was believed by some that brass horns would give a more natural tone to brass instrument recordings. As written in the October 1899 issue of the *Gramophone Record*, "A special curved brass horn has just been placed upon the market... It is well known that the tone of a cornet, or a similar instrument, is only produced after having traversed a certain length of peculiarly curved tubing... Our laboratory has just produced a horn for the Gramophone, which has this principle as its base, and one can fancy the effect, as a Gramophone sound box is placed upon the horn of the usual shaped band instrument in order to reproduce a band record." *Courtesy of Carol A. Veome.*

532 MONTGOMERY WARD & CO.'S CATALOGUE No. 67. 1901

Record Discs.

The Gem Carrying Case.

New Brass Concert Horn.

The "Exhibition" Sound-Box.

Gram-o-phone Record Cabinet.

No. 50 Gram-o-phone Record File.

The Edison Gem Phonograph.

The Edison Standard Phonograph, Complete, $20.00.

Standard Outfit No. 29.

Standard Outfit No. 48.

Edison New Standard Phonograph.

You can hear the latest opera as you sit quietly at home reading your evening paper, if you own an Edison Phonograph. It plays 1,000 tunes. You can make records of your own voice, too. More fun than any game. Produces the same results as the other famous models of the genuine Edison Phonograph. Makes records, reproduces records. Equipped with shaving device. Simplest, most durable, and cheapest talking machine. See No. L 2242, page 532.

1-5. This page from a Montgomery Ward & Company catalogue, prior to May 1900, offered several interesting accessories, including the Leforestier Gramophone horn for $10.00. Note also the engraving of the Edison "Gem" Phonograph (see next illustration). *Courtesy of Ray Phillips.*

1-6. The earliest renderings of the Edison "Gem" Phonograph (see previous illustration) show it with a two-piece conical horn. It is open to question whether the first "Gems" were actually equipped with such a horn, or if the cut displays artistic license. Shown here is a horn, measuring 6" x 14", that roughly resembled the catalogue illustration. *Courtesy of Jen and Ken Flaherty.* (Value code: C)

1-7. A similar conical horn, upon close examination, reveals the May 22, 1905 patent date of the Tea Tray Company. A factory-made extension was appended that increased its length by four inches. It's conceivable that large stocks of 10" long cone horns remained unsold when they became virtually obsolete following the turn-of-the-twentieth-century. This would have represented a smart way of "updating" them. *Courtesy of the Domenic DiBernardo collection.* (Value code: C)

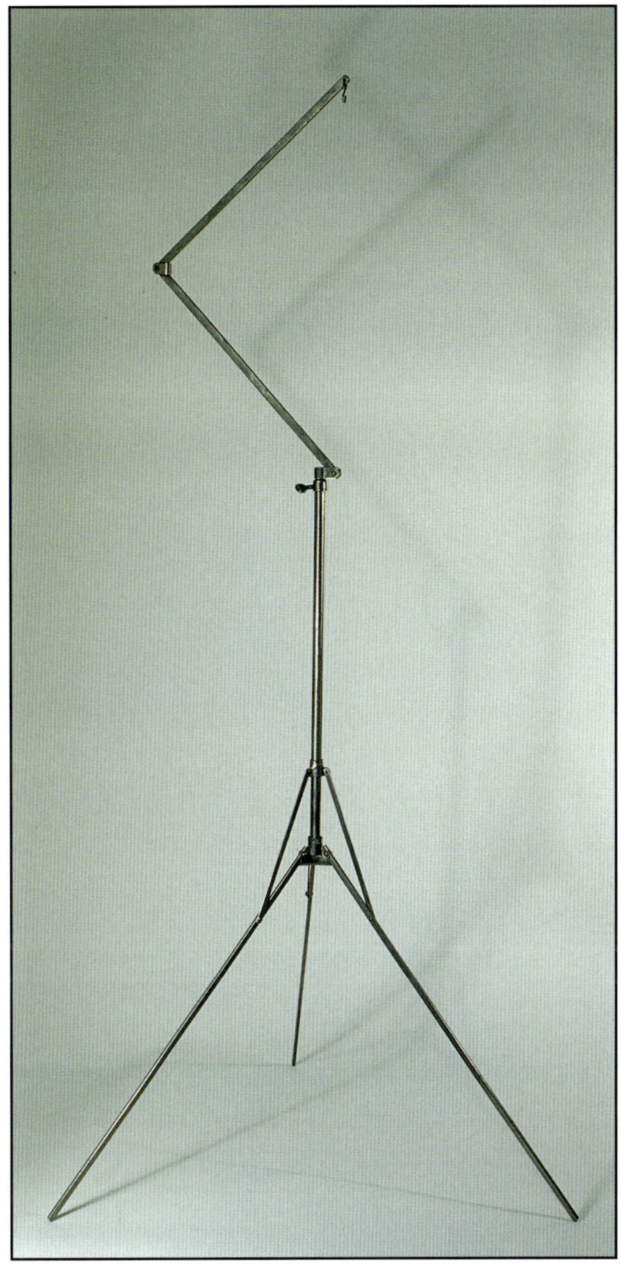

1-8. The simplest solution to suspending a larger cylinder talking machine horn in the 1890s was a floor-standing tripod support, based upon existing technology such as collapsible music stands. As cabinet-mounted horn supports or cranes grew more popular after 1900, floor stands such as this one fell from favor and were used only for horns of lengths greater than 30". *Courtesy of the Scott and Denise Corbett collection*. (Value code: D)

1-9. Tabletop stands to support horns of 14" to 18" length were not commonplace. It seems that most talking machine owners must have opted for horns of larger size, requiring instrument-mounted supports or tripod floor stands. Here we see a 14" brass horn suspended from a Columbia-type table stand. *Courtesy of Pat and Jamie Jones.* (Value code: D)

1-10. A tripod floor crane – in miniature. Note the diminutive size of this stand in relation to a Columbia "Eagle" Graphophone. Unlike other table mounted supports which differed greatly from their full-sized cousins, the one shown here is "anatomically correct" in every way except size. (Value code: VR)

1-11. This horn support was widely offered circa 1900 to fit the Edison "Home" Phonograph. Hawthorne and Sheble sold it, as did the Greater New York Phonograph Company (for $1.50). In this instance, the cabinet of the machine has been embellished with an H&S advertising decal. *Courtesy of Pat and Jamie Jones.* (Value code: E)

1-12. An unusual support arm extension and stabilizing foot for the early (circa 1900) Type "A" through "D" Zonophones. This accessory would enable the substitution of a 30" long horn for the smaller horn supplied with these machines. Shown mounted on a Type "A" Zonophone. *Courtesy of the collection of Howard Hazelcorn.* (Value code: VR)

1-13. Lightness was a strict necessity in a horn designed to be supported by a Bettini reproducer. Key features of a Bettini horn were its significant flare and expanded "neck." Anything but the most featherweight materials would have created too much inertia for the phonograph motor to overcome. Aluminum and zinc were utilized, but papier mâché, as shown here painted in Bettini's signature Chinese red, was also employed. 10" diameter bell, 18" long. *Courtesy of the Domenic DiBernardo collection.* (Value code: A)

1-14. A Pathé-Bettini attachment with a striking red Bettini papier mâché horn on a Pathé "Le Coq," closely modeled on the Type "B" Graphophone. The reproducer is of Pathé manufacture, based upon the Bettini design. *Courtesy of the Julien Anton collection*. (Value code: A)

1-15. Some horns cry out for special handling. Shown is an 18" long glass horn with its carrying case. *Courtesy of Jalal and Charlotte Aro*. (Value code: VR)

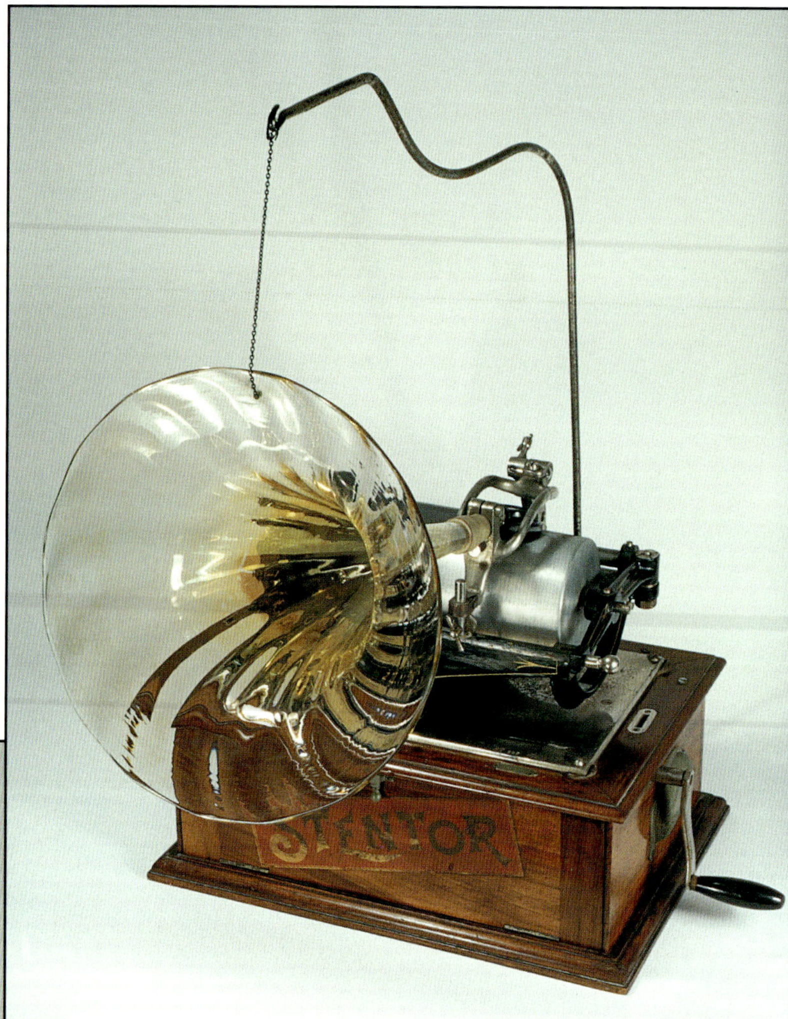

1-16. The 18" glass horn from the previous illustration is shown here mounted on a Stentor ("loud-voiced") phonograph. *Courtesy of Jalal and Charlotte Aro*. (Value code: VR)

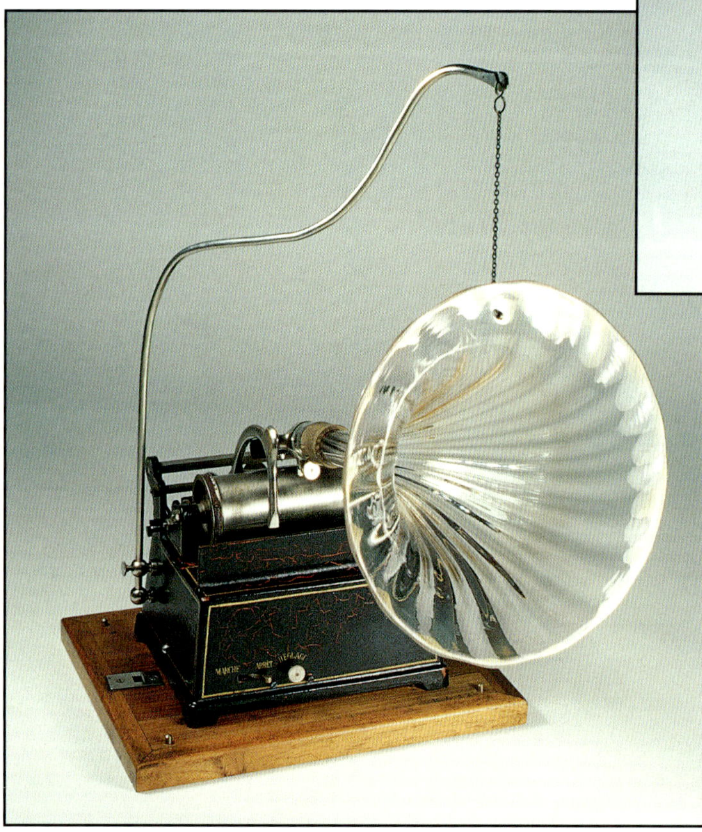

1-17. This first model of the Pathé "Le Ménestrel" (1901) is equipped with a pale pink fluted glass horn measuring 8 3/4" diameter x 11" long. Note that the machine is a factory "paint over." The earlier red finish is discernable through age cracks in the dark blue paint. *Courtesy of the collection of Howard Hazelcorn. (Value* code: VR)

1-18. Mounted on a massive Pathé "Le Céleste" is a 16" long transparent glass horn. One advantage of this horn is *clear* – it does not obscure the beauty of this formidable instrument. *Courtesy of Jalal and Charlotte Aro. (Value* code: VR)

1-19. When talking machines were transported, the horn was often slipped into a canvas bag or even carried unprotected, while the machine was usually cocooned in a leather or wooden case. Here is an exception to that rule: a European Zonophone whose horn fits into a wicker basket, complete with handle. (Value code: C, horn & basket only.)

1-20. Somewhat more conventional is this 12 1/2" diameter x 22" nickel-plated brass horn and traveling arm (for a British Gramophone) with its original flannel-lined carrying bag secured with leather straps. *Courtesy of Jalal and Charlotte Aro.* (Value code: C)

1-21. Of all the materials used to create early disc talking machine horns – brass, steel or zinc – papier mâché would exceed them all in fragility. For this reason, it's exciting to find one such delicate artifact preserved after a century of risk. *Courtesy of the Domenic DiBernardo collection.* (Value code: C)

1-22. The *cor de chasse* or "hunting horn" was a European style that was not imitated in America (except for the "Wizard" phonograph of 1911), despite its advantages of size, space, and self-support. This Pathé "Le Sublime" is equipped with one of these efficient horns made of aluminum. *Courtesy of the collection of Howard Hazelcorn.* (*Value* code: D)

1-23. Another view of a *cor de chasse*, this one in brass and marked "Pathé." The efficiency of space offered by these horns is evident, and would be emulated in 1909 with the introduction of the Edison "Cygnet" horn. *Courtesy of Garry James*. (Value code: C)

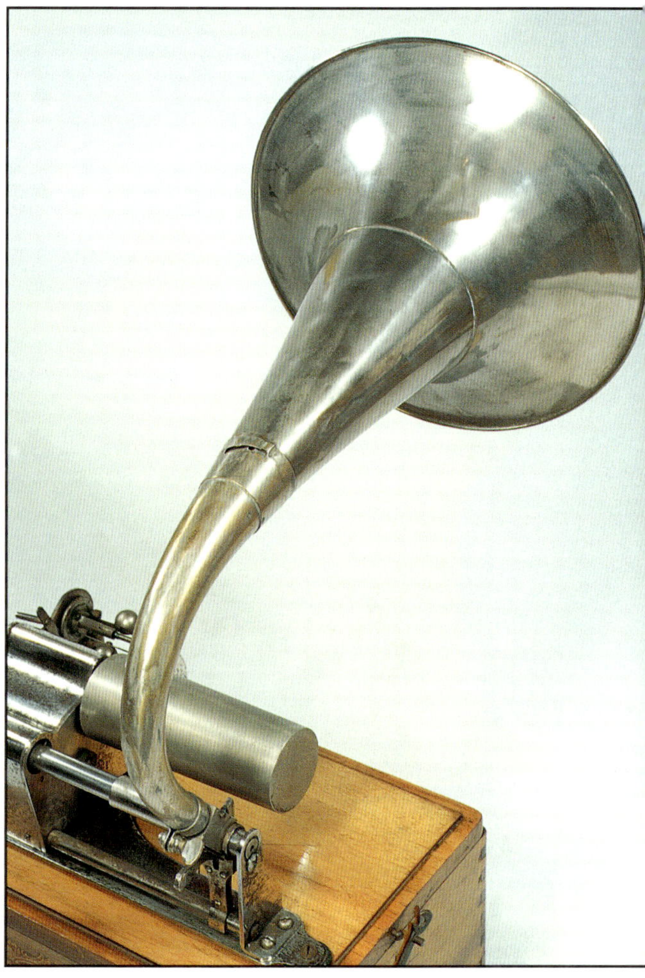

1-24. If a customer could not decide between a straight horn and a *cor de chasse*, perhaps this convertible horn would do. The nickel-plated horn is shown here with its *cor de chasse* elbow (see next illustration). *Courtesy of Yves Rouchaleau* (Value code: D)

1-25.The same horn, shown with its straight elbow attached, results in a much shorter sound passage. *Courtesy of Yves Rouchaleau*

1-26. This 31" long European nickel-plated horn for a cylinder phonograph is in a most unusual pasteboard carrier. *Courtesy of Jalal and Charlotte Aro*. (Value code: C)

1-27. Ear-tubes survive in limited numbers, but century-old headphones are quite rare. Shown is a close-up of a set marked, "H.E. Booth, CLARAPHONE, Pat. Sept. 15, [19]03. Made by Columbia Phonograph Company." Devices like this were probably used most often with commercial Graphophones. (Value code: VR)

1-28. Circa 1901-1905, the Tea Tray Company, Newark, New Jersey, marketed a popular line of straight horns bearing a red label declaring, "ACCEPTED STANDARD AMPLIFYING HORN FOR TALKING MACHINES." Many of these horns were bought by the owners of cylinder phonographs, who found the 14" long horns supplied with their instruments inadequate. It is unusual to find one of these horns manufactured for a Victor talking machine. This example has a threaded end to accept a Victor elbow, and lacks a hanger for attaching to a cylinder machine crane. This horn would have required a specially extended Victor support arm and carrier arm. Such a product would have appealed only to the rare Victor owner whose craving for a larger horn was not satisfied by the Victor catalogue. Measures 13 1/2" diameter x 27 1/2" long. *Courtesy of the Babcock House Museum, John H. Perschbacher*. (Value code: D)

1-29. The Victor "VI" (introduced 1904) was first equipped with a black fiber horn, which was quickly replaced by a mahogany model (No. "31") or a brass flower horn. Shown is a rare example of the early metal mahogany-grained horn supplied with the Victor "VI" for a short period before the wooden and flower horns became available. *Courtesy of Brad Abel*. (Value code: VR)

1-30. Pathé presented a field of phonographs rife with unusual methods for supporting a horn. A number of interesting attachments were appended to Pathé machines. Shown is the "Le Français" with a "Gallier" attachment, looking as though it belonged on a disc talking machine rather than a cylinder instrument. (Value code: B)

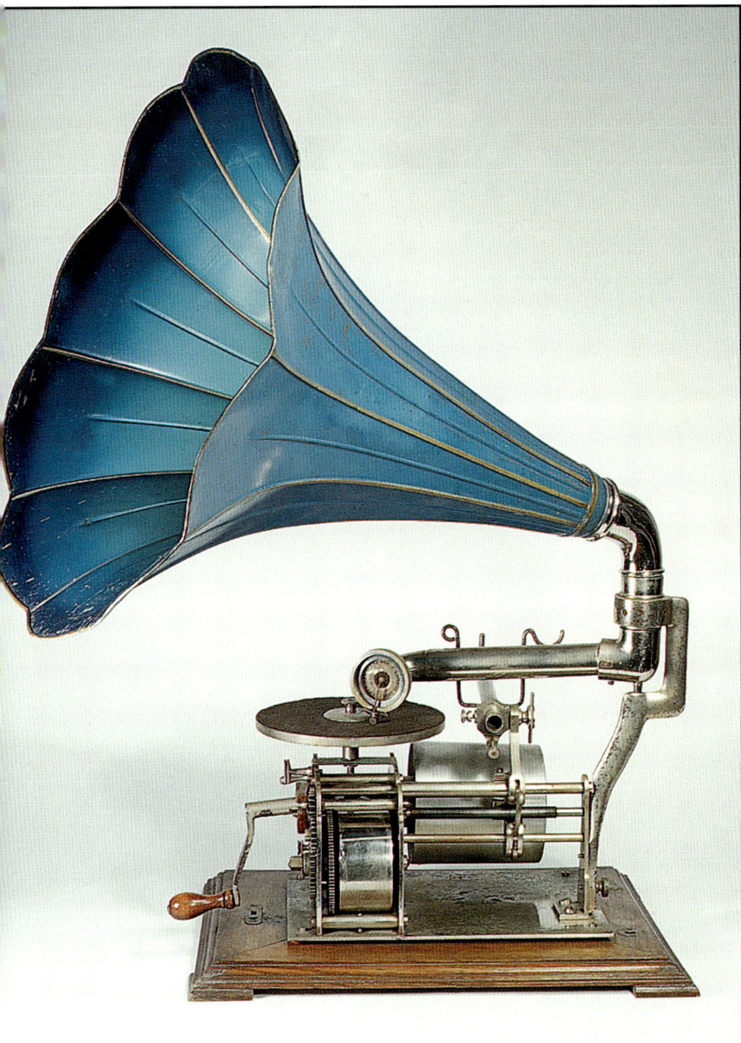

1-31. This combination disc/cylinder Idéal "No.5" is breathtaking, but it's the back-mounted horn in shades of blue that catches the eye. Earlier, more frequently seen versions of this machine employed a front-mounted horn of aluminum, not approaching the visual impact of this example. *Courtesy of the Julien Anton collection*. (Value code: VR)

1-32. Some horns were designed for looks, some for sound, a ew probably for both. This striking aluminum horn measuring 13 1/2" diameter x 16 1/2" long features concentric tiers, which may have been touted as acoustically superior, but certainly made for a visually interesting accessory. The machine is a Bettini "No.4" with a conventional carriage, fitted with an ebonite "Le Cahit" reproducer. The crane isplays a graceful design typical of European horn supports. *Courtesy of Richard and Jill Pope.* (Value code: VR)

1-33. The Standard Metal Manufacturing Company of Newark, New Jersey was a well-known supplier of quality horns. As trumpet-shaped horns gave way to flower-shaped designs during 1904-1905, Standard is the only known manufacturer to have offered decorated trumpet styles, such as the one shown. Measures 16 3/4" diameter x 30" long. *Courtesy of Jen and Ken Flaherty*. (Value code: C)

1-34. Another decorated Standard horn of the same dimensions, this one with daisies. *Courtesy of Pat and Jamie Jones.* (Value code: C)

1-35. The Standard Metal Manufacturing Company briefly offered this very unusual finish, described as "crystal and gold" in an April 1905 advertisement in the *Talking Machine World*. Measures 16" diameter x 31" long. *Courtesy of Tom Sideleau.* (Value code: C)

1-36. Most unusual are "seasonal" themes, which would have been out of place for most of the year. This 16 3/4"diameter x 30" long Standard horn features Christmas holly. *Courtesy of Robin and Joan Rolfs.* (Value code: C)

1-37. This odd horn crane by the Standard Metal Manufacturing Company literally embraced the phonograph to which it was attached. As could be expected, the use of thumbscrews to secure crane to cabinet often marred the finish. Note the substantial anti-tip foot in the foreground. The crane bracket listed for $1.90, while the extension crane was $.75. The machine is an Edison "Standard" Model "A" with an International Correspondence Schools repeating attachment for language study, which would normally not require a large horn. *Courtesy of Howard Embleton*. (Value code: D)

1-38. This unusual crane socket was designed to be secured in the lid latch escutcheons found on most Edison Phonographs. Shown partially inserted to reveal the securing slot. *Courtesy of Howard Embleton*. (Value code: D)

1-39. $1.25 bought you an "O.K. Horn Crane," made by the Ideal Fastener Company. As an advertisement in the September 1906 *Talking Machine World* gushed, "The Acme of Simplicity and Perfection... (Instantly attached by the turn of a thumb nut)... For Edison and Columbia Machines." The "O.K." is shown here attached to a Columbia "BK" Graphophone, but it would work on any instrument having slots in the motor mounting plate to accept lid fastening arms. *Courtesy of the Scott and Denise Corbett collection*. (Value code: D)

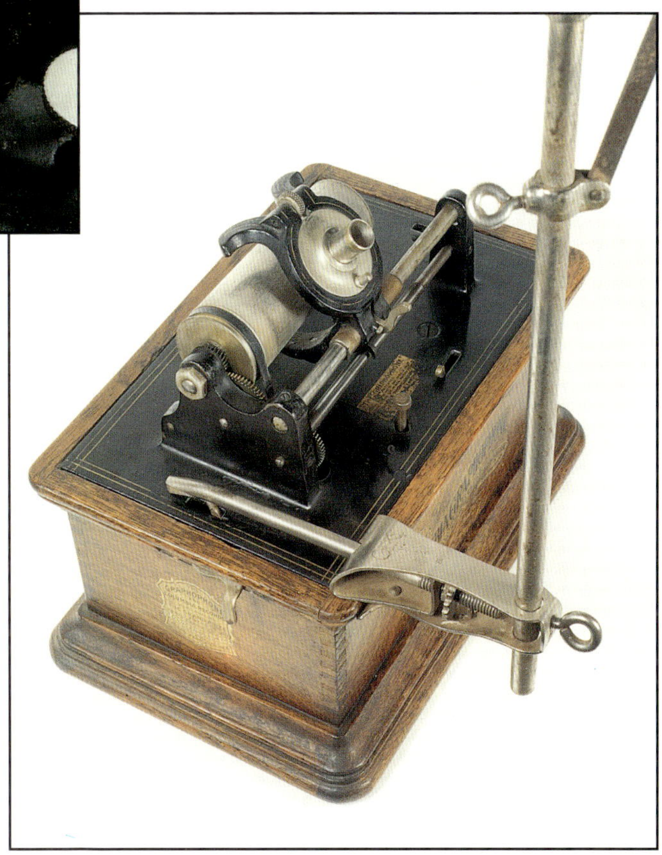

1-40. Gazing at the "Chicago Horn Support For Edison Phonographs" one is struck by how complicated it is – seemingly at the expense of durability. Compare it, for instance, to the sturdy device in illustration 1-42. The Chicago Stand Company provided insight regarding the unusual design in an advertisement that appeared in the August 1906 *Talking Machine World*, "The support is constructed so that it comes directly under the machine and practically out of sight, leaving the top of the machine perfectly clear." In other words, no clunky arms to get in the way. The support's fragile profile, however, might not have elicited confidence from a prospective buyer. It retailed for $1.50, wholesaled for 75 cents – an unusually good profit for the merchant. *Courtesy of the Scott and Denise Corbett collection.* (Value code: D)

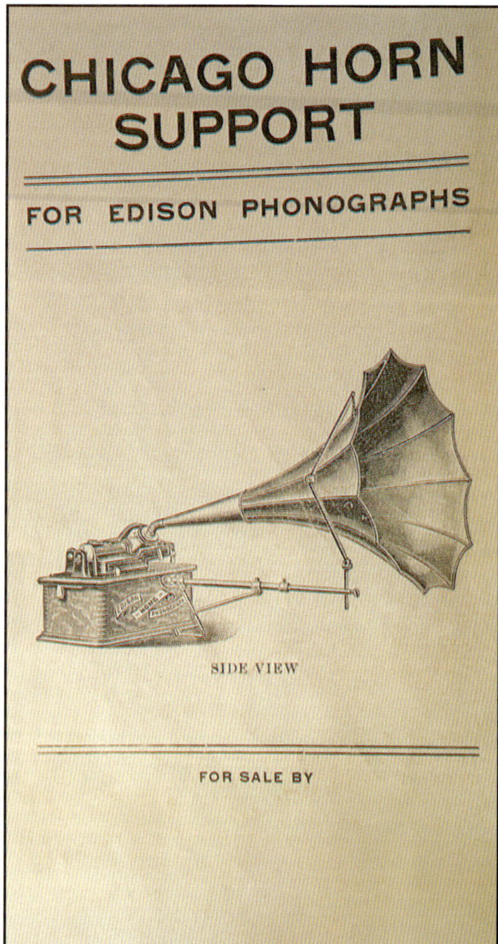

1-41. A sales flyer illustrating the Chicago horn support. *Courtesy of Michael F. Khanchalian.*

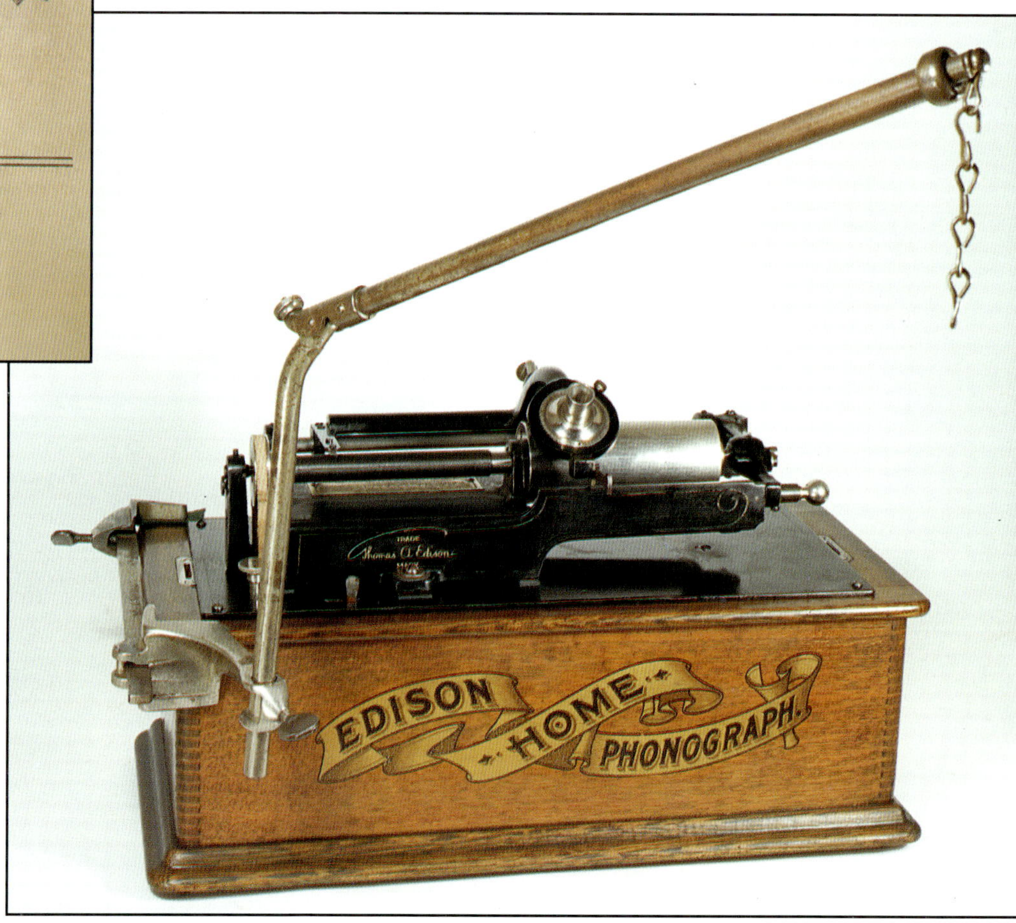

1-42. The Ideal Fastener Company of New York City produced this clever horn support which it first advertised in the May 1906 *Talking Machine World*, "The Universal Horn Crane. A crane **equally well adapted** for **Edison Phonographs** and **Columbia Graphophones.** A novel, effective device for raising or lowering horn. Fully nickel-plated. …At all jobbers… Retail Price - - $1.75, Dealer's Price — $1.00." This crane works equally well in the position shown here on the upper molding, or lower on the body of the cabinet. (Value code: D)

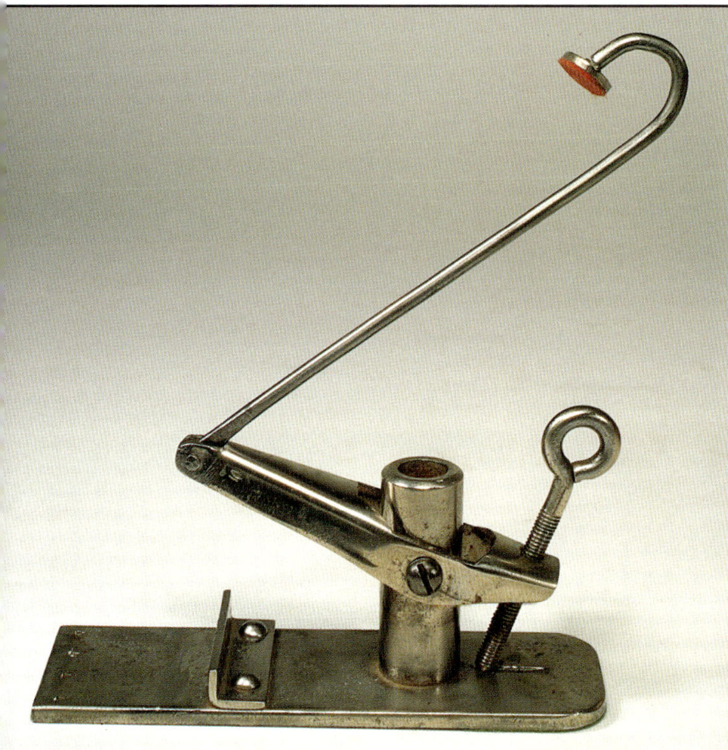

1-43. Some horn support companies went to considerable lengths to secure their products in place – especially to avoid drilling holes in or otherwise marring the cabinet of the talking machine. The base of this support is an excellent example of stylish ingenuity. The "foot" slips under the base cabinet molding, and the graceful "finger" reaches up to press down upon the upper surface of the instrument. Tightening the eyebolt makes the support snug, with no damage to the woodwork. *Courtesy of Jean-Paul Agnard.* (Value code: D)

1-44. Horn support manufacturers applied a great deal of ingenuity to creating convenient designs. Here, the base of the support was secured to the machine by the same latching mechanism into which the lid locked. This obviated the need to drill holes or otherwise alter the cabinet, but precluded the use of the lid as a dust cover while the crane was in place. (Value code: D)

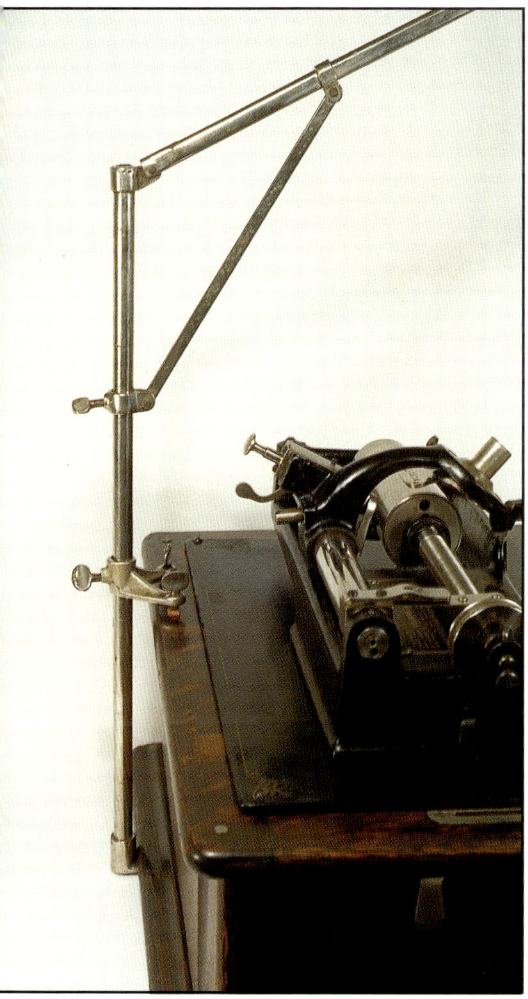

1-45. This horn crane affixes to the rear of a cylinder talking machine cabinet by the use of two brutal thumbscrews, protected here from distressing the wood by small rubber pads. (Value code: D)

1-46. An unmarked fiber horn with Japanese lacquer decorated finish. These horns were sometimes advertised generically as "Kyoto" horns. *Courtesy of Pat and Jamie Jones*. (Value code: A)

1-47. One of the better-known brands of oriental-style horns was the Allen's Paper Lacquer Horn. The sole distributor for the United States was Byron Mauzy of San Francisco. This well-preserved example is mounted on a Victor "III." *Courtesy of Fred Guenther.* (Value code: VR)

1-48. The marking on the Allen's Paper Lacquer Horn. The foreign patent No.1022 is sometimes found on the Victor fiber horns that were supplied with the earliest Victor "VI." *Courtesy of Fred Guenther.*

1-49. Another Allen's Paper Lacquer Horn, this one with a painted Japanese landscape. This example bears the same marking as seen in the previous illustration. *Courtesy of Robin and Joan Rolfs*. (Value code: A)

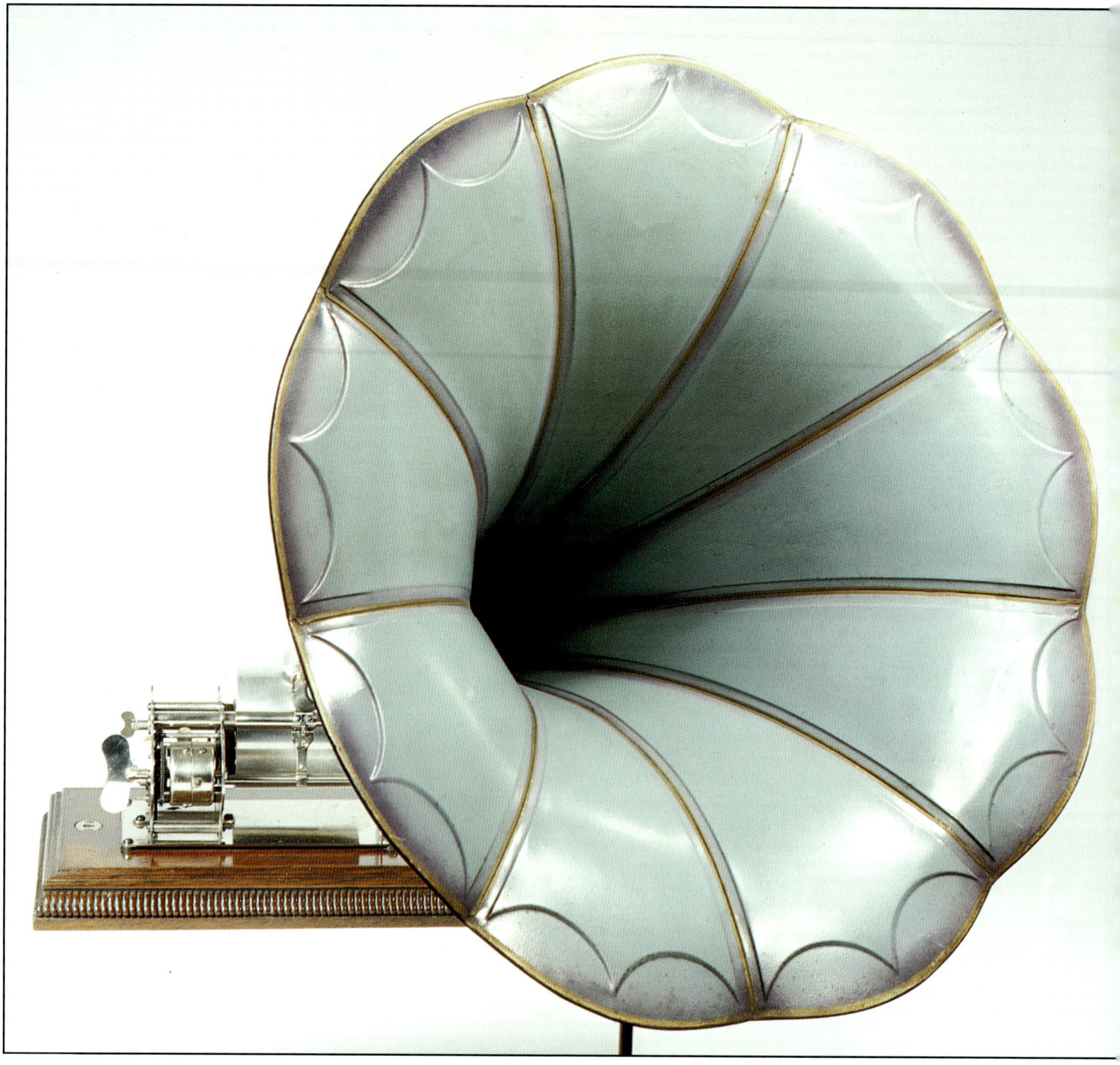

1-50. A 22" diameter x 30" long European flower horn in delicate pastel shades is mounted on a Type "AB" Graphophone. Flower horns of this size are highly unusual in Europe. It appears that most Continental patrons, probably due to domestic space limitations, chose horns that would have been considered small in the United States. *Courtesy of the Julien Anton collection*. (Value code: VR)

1-51.A spectacular paneled horn with mixed floral decoration, surmounting a Victor "MS" talking machine. *Courtesy of the Domenic DiBernardo collection.* (Value code: B)

1-52. This small red "Mega" Phono Horn measures 17" diameter x 25 1/4" long. E.A. and A.G. Schoettel of Maspeth, New York, manufactured these lightweight fiber horns, advertised in the *Talking Machine World* from January 1905. By November of that year, "flower"-shaped horns such as this one were available. Surviving "Mega" horns are usually larger than the one shown here. *Courtesy of Charles McCarn*. (Value code: D)

1-53. A larger red "Mega" Phono Horn featuring boldly executed white mums. Decorated "flower" Mega horns were available by July 1906. Measures 23" diameter x 30 1/2" long. *Courtesy of the Gfell family collection*. (Value code: C)

1-54. A "Mega" horn of unusual color and size. This gilt example with floral decoration measures a diminutive 12 3/4" diameter x 22" long. *Courtesy of Jean-Paul Agnard*. (Value code: D)

1-55. This violet colored "Mega" horn with pansies was slightly shortened by a previous owner. It currently measures 23" diameter x 30" long. *Courtesy of Jean-Paul Agnard*. (Value code: D)

1-56. Still another "Mega" Phono Horn, this one in black with delicately painted flowers. The last advertisement for decorated "Mega" horns appeared in the *Talking Machine World* in November 1907. *Courtesy of Jean-Paul Agnard*. (Value code: D)

1-58. This unusually short and squat (6 1/8" diameter x 10" long) horn may have been original equipment on certain client-goods machines (sold by retailers under their own brand names), such as this unmarked Columbia product. The horn bears the patent date of August 22, 1905, which was controlled by the Tea Tray Company. *Courtesy of the Scott and Denise Corbett collection.* (Value code: E)

Opposite page: 1-57. The Talkophone Company of Toledo, Ohio held onto its position as an independent in the talking machine industry by the skin of its teeth. During the first decade of the twentieth century, the firm was sued for patent violation by both Victor and Columbia. Though it would wither under legal and financial onslaughts, Talkophone managed to make the transition from "front-mounted" to "back-mounted" design. Evidence indicates that the same type of horn was employed for all Talkophone back-mounted instruments: a flower horn painted with "morning glory" effect, proceeding from yellow at the innermost portion to a deep purple at the extremities of the panels. Only the size of the horn was varied to fit various models; the appearance remained the same. In this instance, the largest version of the Talkophone horn is shown mounted on a later version of the fancy "Sousa." Since Talkophone used an elbow fitting identical to Victor's, it is possible to find Talkophones today incorrectly equipped with horns intended for a Victor. *Courtesy of Greg and Jane Rush.* (Value code: D)

1-59. The oddly shaped Search light Company horns ("Design and Constructed on the Scienti Principle of a Searchlight Reflector by a Master of Acous tics") for cylinder or disc machines appeared in June 190 and were usually painted or plated. With very few exceptio neither treatment has satisfacto rily withstood the passage of time. This is an unusual examp in brass, measuring 18 1/2" diameter x 19 3/4" long. Manufactured by the Searchlight Ho Company, Brooklyn, New York *Courtesy of Jerry Blais (horn only* (Value code: VR)

1-60. This Searchlight horn for a cylinder phonograph features original decoration of yellow mums. Unfortunately the deep purple aniline dye in the finish faded away during a century of exposure to daylight. A flicker of its regal beauty may be discerned in the deepest interior of the horn. Measures 20 3/4" x 31". *Courtesy of Garry James*. (Value code: D)

1-61. What might appear at first glance to be an ordinary Searchlight horn (with typical aged finish) is actually the company's first model collapsible horn, announced in December 1906. *Courtesy of the Gfell family collection.* (Value code: VR)

1-62. The first model Searchlight collapsible (or "Knockdown") horn showing a section removed. In order to assemble or take down this horn, 12 thumbscrews had to be negotiated, along with six small but devilish clips located on the perimeter of the bell. *Courtesy of the Gfell family collection.*

1-63. The second model of the Searchlight collapsible horn appeared in August 1907. Although more conventional than its predecessor, the folding, piano-hinged petals were not a recipe for scratch prevention. This example retains a remarkably well-preserved finish. First offered in "Gun Metal Finish," by October 1907 the horn was available in "Red (Baked Japan)" as shown. Measures 23" diameter x 30 1/2" long. *Courtesy of the Babcock House Museum, John H. Perschbacher.* (Value code: VR)

1-64. The second model Searchlight collapsible horn broken down into its component parts. The "interior" cone at top is inserted through the expanded body of the horn, into the "exterior" cone (painted red). Then the threaded retaining ring is screwed in place, drawing the parts together. *Courtesy of the Babcock House Museum, John H. Perschbacher*.

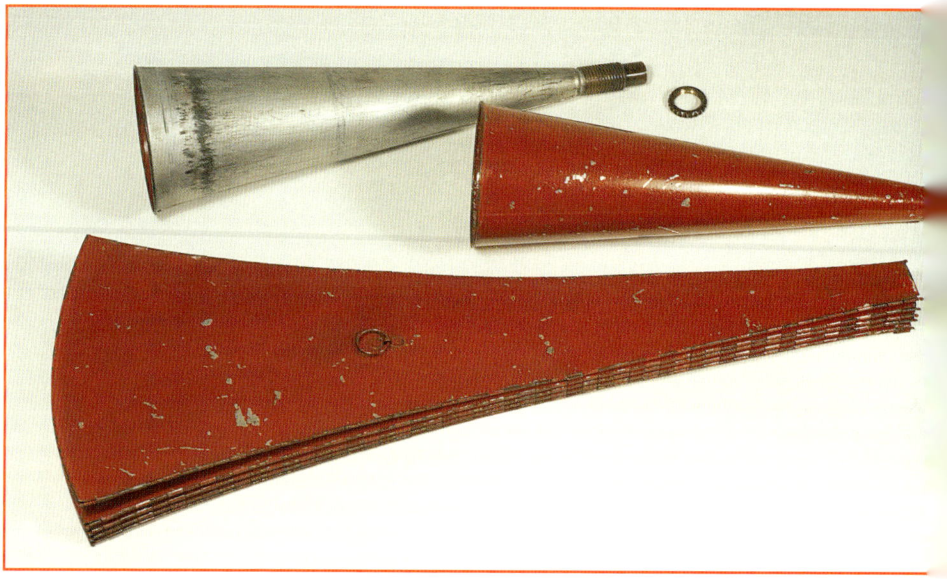

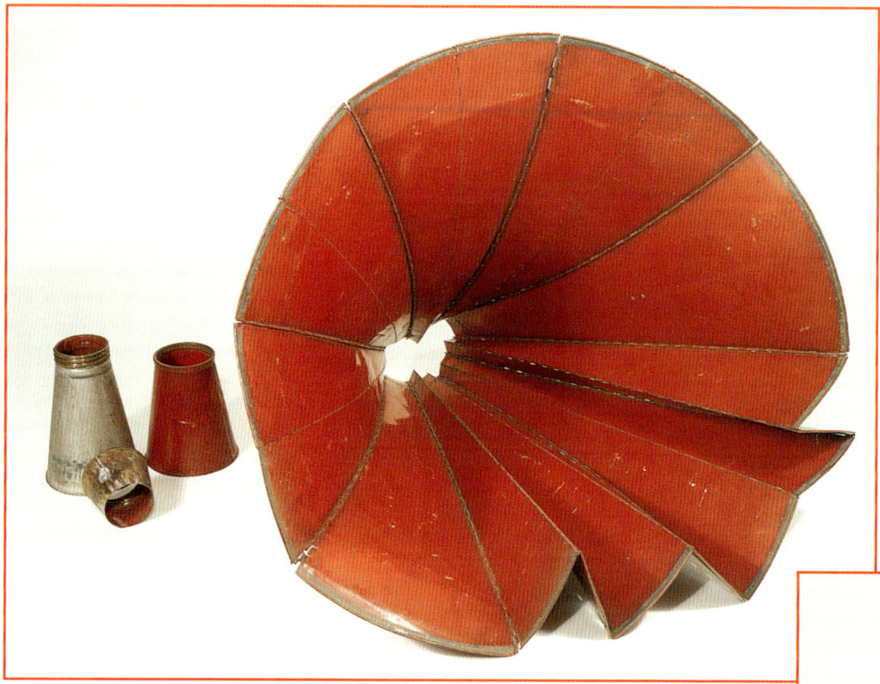

1-65. The disc model of a Searchlight collapsible horn is shown here partially assembled. An extra collar screwed onto the Columbia-type threads adapts the horn for a Victor machine. *Courtesy of John T. Hoffman, the Phonophile*. (Value code: VR)

1-66. Detail of the disc Searchlight collapsible horn connection. In practice, screwing this complex assembly into a Columbia elbow must have been tricky. By June 1908, the Standard Metal Manufacturing Company, Newark, New Jersey, announced that, "We have taken over the ENTIRE PLANT of the Searchlight Horn Co. All future business in this line will be taken care of by us." *Courtesy of John T. Hoffman, the Phonophile*.

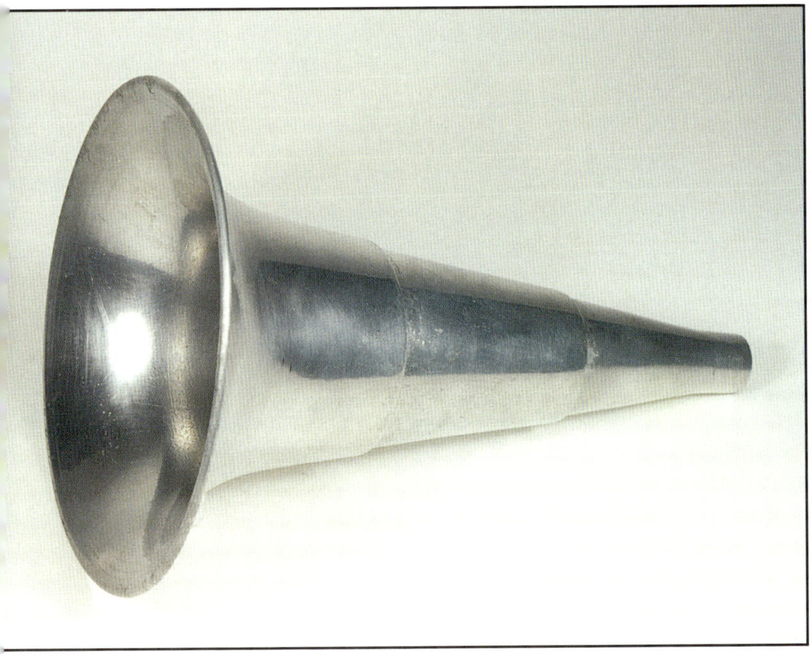

1-67. This small, collapsible aluminum horn is unmarked. Measures 7 3/4" long assembled. *Courtesy of the Charles Hummel collections*. (Value code: VR)

1-68. The previous example shown broken down into three pieces. Such a diminutive horn could almost be carried in a pocket! *Courtesy of the Charles Hummel collections*.

1-69. One of the better-promoted collapsible horns was the Munson Folding Horn, advertised in the *Talking Machine World* from October 1907. First offered for cylinder phonographs, the Munson became available for disc machines in April 1908. After November 1909, no further advertising for this horn appeared in the *Talking Machine World*. As might be expected, these fabric horns often do not age gracefully, so this nice example with pansy decoration is a particular treat. Bell is 21" diameter. Manufactured by the Folding Phonographic Horn Company of New York. *Courtesy of Bernie and Katie VonThaer*. (Value code: VR)

1-70. Victor H. Rapke of New York City manufa tured talking machine accessories, including the "Rapke Horn" shown here. The concept of a "collapsible" horn for cylinder machines had be explored by Columbia, which marketed a celluloid-bodied horn in various sizes. The Columbia horn could be stored nearly flat because the met bell could be detached and the flexible body depressed. Rapke explored a design that involve aluminum sections of expanding size snapped together to produce a horn that could be carried either by the reproducer (as seen here) or of sufficient length to require the support of the "Rapke Crane." An advertisement in the January 1907 *Talking Machine World* stated, "The horn reproduces the record naturally without that disagreeable metallic tone, and is entirely free of foreign noises, vibration or 'rattle.' Impossible to get out of order. Any person can set it up. Can be packed in small compass and sent through the mails. Can be put together or taken apart in a second. Locked or unlocked, this horn will susta the weight of any man standing upon it. Can be used on any machine: Edison, Victor, Columbia, Twentieth Century, Zonophone, Auxetophone an all others." The addition of certain machines in this list proves surprising. The advertisement illustrates a Rapke horn like the one pictured her affixed to an Edison Home cylinder Phonograph. However, the text indicates that various alteratio could be made to the sections to allow it to be use with disc talking machines. This makes sense, an underscores the value of the Rapke design – to change between configurations, the substitution c one or two of the demountable sections was all th was required. (Value code: VR)

1-71. The General Phonograph Company of New York sold "Truetone" Amplifying Horns and first advertised the fact in the November 1907 issue of the *Talking Machine World*. These horns were made of thin wooden petals glued to shaped ribs. Although lighter in weight than most other wooden horns, they were (and remain) susceptible to warping due to dampness and atmospheric changes. The "Medium Size" horn with a 19" bell for cylinder or disc machines was $6.00. "Large Size" with a 23" bell for cylinder or disc was $7.50. *Courtesy of Norm and Janyne Smith*. (Value code: A)

1-72. An unmarked purple flower horn with pink blossoms. Measures 23" diameter x 31" long. *Courtesy of Dan and Sandy Krygier*. (Value code: D)

1-73. This iridescent green flower horn features pink roses along its entire perimeter. Unmarked, it measures 23" diameter x 31" long. *Courtesy of Dan and Sandy Krygier*. (Value code: C)

1-74. A Type "BKT" Graphophone sports a blue flower horn with pink roses. Columbia itself did not catalogue horns with painted floral decoration, and extant examples are unusual. *Courtesy of Charles McCarn.* (Value code: D)

1-75. Unusual in its use of bouquets on the interior, this deep blue horn measures 22" diameter x 30" long. *Courtesy of the Gfell family collection*. (Value code: C)

1-76. Pink and yellow roses decorate the interior of this unmarked red flower horn. Measures 23" diameter x 31 1/2" long. *Courtesy of the Gfell family collection.* (Value code: D)

1-77. Dogwood blossoms decorate this unmarked horn painted in unusual teal blue. Measures 22" diameter x 31" long. *Courtesy of Richard and Nancyann Brown.* (Value code: D)

1-78. Pansies decorate the interior of this green horn manufactured by the Tea Tray Company of Newark, New Jersey. Measures 22 1/2" diameter x 30 1/2". *Courtesy of Skinner's Auctions.* (Value code: C)

1-79. Deeply scalloped petals are characteristic of Tea Tray Company horns (although the company marketed other designs as well). Measuring 22" diameter x 30 1/2", this green example sits atop a mahogany Edison "Triumph" Phonograph Model "B." *Courtesy of Pat and Jamie Jones*. (Value code: C)

1-80. A red horn of the Tea Tray Company, fitted to a Victor "II." Note that the pink and yellow pansy decorations flow across the panels in the conventional manner (see next illustration). Measures 22 1/2" diameter x 24" long. *Courtesy of Charles McCarn*. (Value code: C)

1-81. This black Tea Tray Company horn mounted on a Victor "III" bears odd floral decoration. Note that the mum on the upper panel is incomplete, suggesting that this horn had been painted and decorated *before* it was assembled. Other more subtle parts of the decoration are missing as well. If this horn is indicative of the results, the "experiment" in pre-painting the panels must have been short-lived. Measures 22 1/2" diameter x 24" long. *Courtesy of Charles McCarn*. (Value code: C)

1-82. This pale blue flower horn is one of the Tea Tray Company's "Art-Tone" models, announced in June 1907. They displayed additional decoration, meant to distinguish them from the many cheaper horns available at the time. "There are no transfers or decalcomanias used in any part or portion of the decoration. It has been estimated by the makers that it requires over five hundred touches of the artist's brush to bring one of these 'Art-Tone' horns up to the maker's ideal." *Courtesy of the Domenic DiBernardo collection.* (Value code: C)

1-83. Another "Art-Tone" horn from the Tea Tray Company of Newark, New Jersey. In additio to the artwork, the unusual strength of these horns was touted: "Our new 'Art-Tone' Flower Horns are so constructec that they cannot be bent or crushed, either in everyday use c in transportation." *Courtesy of th Domenic DiBernardo collection.* (Value code: C)

1-84. An unmarked black horn with a spray of pink roses dresses up a Victor "III." Horn measures 22" diameter x 22 3/4" long. *Courtesy of Nina and Dave Heitz*. (Value code: C)

1-85. This Victor "V" is distinguished by an unmarked gold-colored horn with pansies. Measures 22 1/2" diameter x 25 1/2" long. *Courtesy of Charles McCarn*. (Value code: C)

1-86. This deep blue eleven-paneled flower horn is striking, with its "wine leaf" decoration. The horn is constructed of brass, and the vine is actually revealing the brightness of the metal beneath the paint. Measures 22 3/4" diameter x 31 1/2" long. *Courtesy of Jean-Paul Agnard*. (Value code: C)

1-87. An unmarked green flower horn with very unusual white roses and ribbon decoration. Measures 23" diameter x 31" long. *Courtesy of Skinner's Auctions*. (Value code: D)

1-88. A brief entrant in the American talking machine horn market was the New Jersey Horn Manufacturing Company, which introduced its "Tulip" horn in July 1907. One advertisement appeared that month in the *Talking Machine World*, after which nothing more was heard of this unusually shaped horn or its manufacturer. This is a prime example of the "Tulip," featuring floral decoration on a base of silver with blue stripes. Measures 21 1/2" diameter x 30 3/4" long. *Courtesy of Brady Jeffcoat*. (Value code: C)

1-89. Production oddities seem to turn up less often with Edison products than with those of Victor or Columbia. Shown is a rare example: a "Gem" Model "B" whose horn was labeled "Triumph!" Other than the erratic decal, this one-piece "Gem" horn is a typical example of its type. *Courtesy of Gregg Cline*. (Value code: VR)

1-90. Another one-piece "Gem" horn: most atypical in that it displays original blue paint. A large jobber of Edison merchandise such as Babson Brothers of Chicago could arrange for Edison horns in non-standard colors, but such straying from the typical, unimaginative Edison black horn equipment was rare. *Courtesy of Jean-Paul Agnard*. (Value code: VR)

1-91. A close-up of the blue 19" Edison horn shows that it is marked neither "Gem" nor "Fireside," the two types of designation seen on this size horn. The mystery begins to unravel, however, when it's revealed that this horn was originally sold in Australia. *Courtesy of Jean-Paul Agnard*.

1-92. The Columbia Phonograph Company had something of a penchant for assembling talking machines from "available" parts – hence later-style cabinets with early motors, and later-style motors in early cabinets. They weren't proud. Even self-important Victor engaged in some "hodge-podging," primarily in the "P" series (Premium). Edison, however, was really too good for this sort of behavior. No mixing and matching for the Wizard of Menlo Park – except when nobody would notice. Certain Edison merchandise, not found in the American catalogue, has turned up in Australia, such as the blue "Gem" horn shown in illustration 1-90. Perhaps the most striking anomaly, however, is this Edison "Triumph" Model "F" (1911) with original sky blue enameled No."11" cygnet horn. Underscoring the peculiarity of this instrument is the fact that the horn bears an Edison "Standard" horn decal instead of the customary cygnet designation. (Value code: VR)

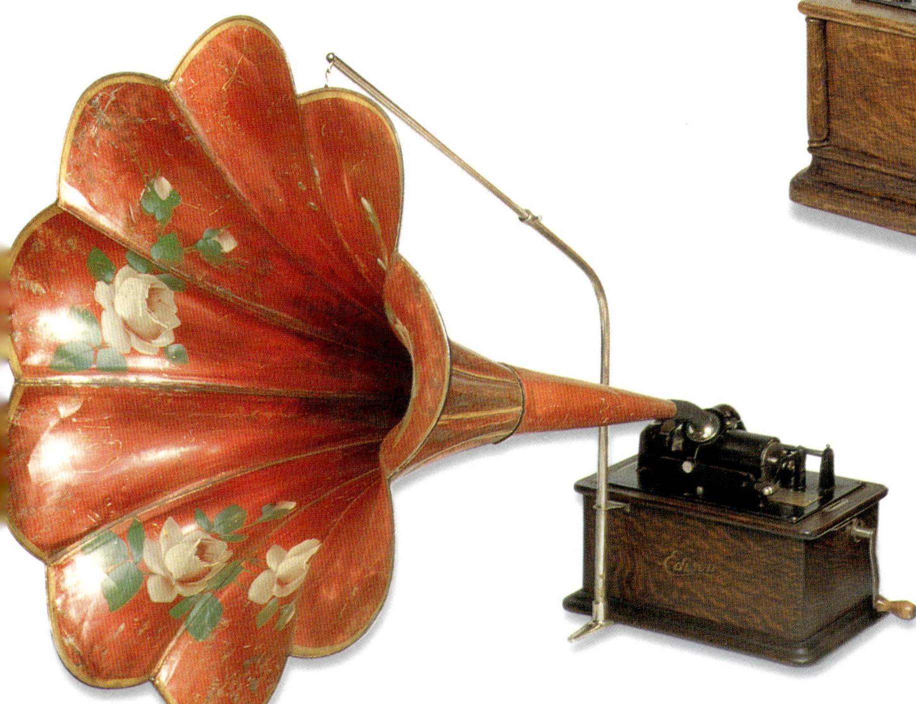

1-93. The Standard Metal Manufacturing Company, in addition to more typical flower horns, introduced its "Lily" horn in June 1908: rather late in the life of the flower style horn. As can be seen, each petal is concave in cross-section, giving this model a distinctive appearance. *Courtesy of Brice Paris (horn) and V.J. Seth (phonograph)*. (Value code: C)

1-94. An august, straight wooden horn with applied brass bell, by the Wooden Phonograph Horn Company of Syracuse, New York. 14" diameter bell, 37" long. *Courtesy of the Domenic DiBernardo collection.* (Value code: A)

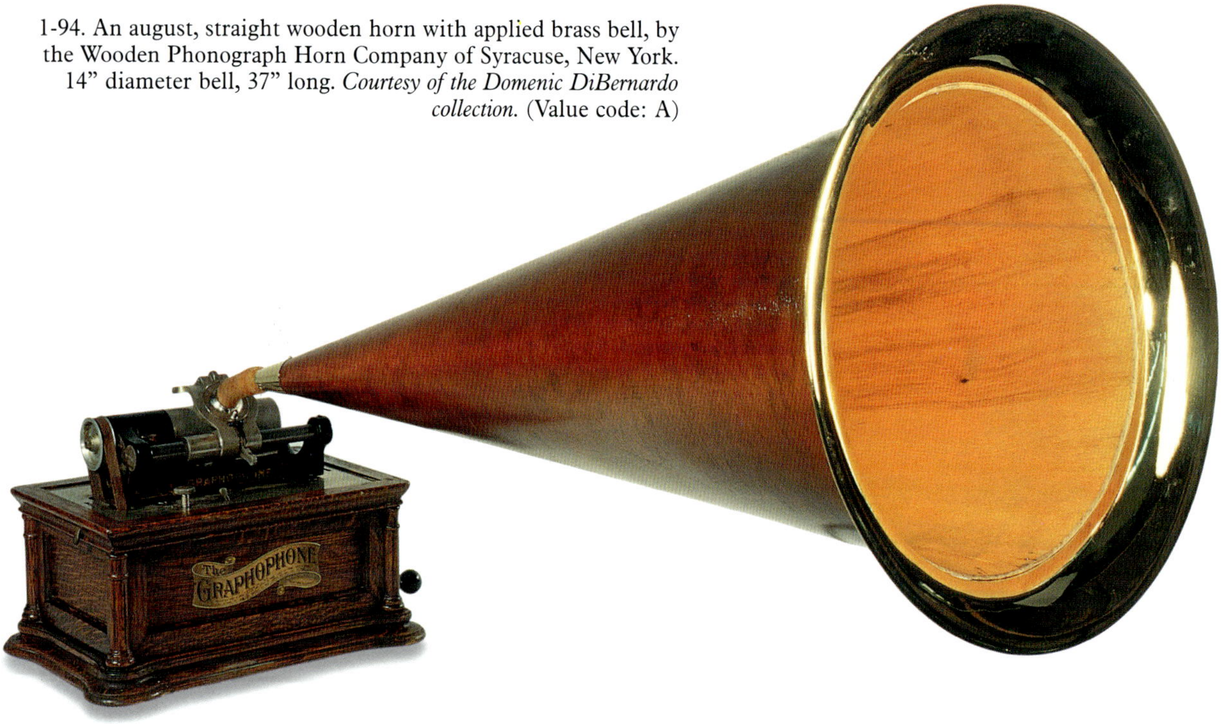

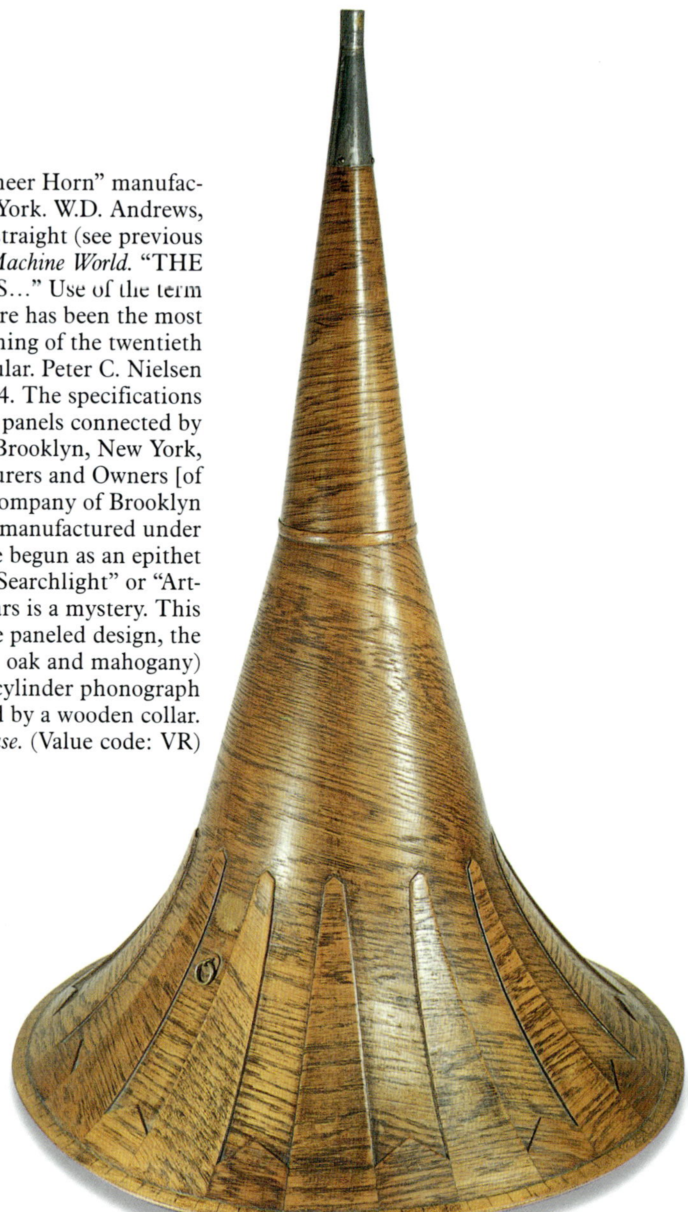

1-95. $10.00 bought you this stunning 22" bell by 32" long "Wood Veneer Horn" manufactured by the Wooden Phonograph Horn Company of Syracuse, New York. W.D. Andrews, sole distributor, with offices in Syracuse and Buffalo, promoted both straight (see previous illustration) and "Morning Glory Style" in the April 1908 *Talking Machine World.* "THE HORN of the hour… FOR CYLINDER AND DISK MACHINES…" Use of the term "Morning Glory" bears noting. Since the 1950s, this nomenclature has been the most common way in America of referring to a paneled horn. At the beginning of the twentieth century, however, the generic phrase "Flower Horn" was far more popular. Peter C. Nielsen was awarded United States patent No.771,441 on October 4, 1904. The specifications governed the structure of talking machine horns with sections or panels connected by longitudinal ribs. The United States Horn Company (Incorporated) of Brooklyn, New York, claimed to be in control of Nielsen's patent, calling itself the "Manufacturers and Owners [of the] MORNING GLORY HORN." In June 1906 the Searchlight Horn Company of Brooklyn announced its eponymous "ribbed" horn in the *Talking Machine World*, manufactured under the Nielsen patent. So, the designation "Morning Glory" seems to have begun as an epithet of no greater importance than others such as "Flower," "Lily," "Tulip," "Searchlight" or "Art-Tone." Why the phrase should have reached such prominence in later years is a mystery. This "Wood Veneer Horn" is very interestingly constructed. In this same paneled design, the Wooden Phonograph Horn Company supplied the No."2" wood horn (in oak and mahogany) that Columbia offered with certain Disc Graphophones. To create the cylinder phonograph version, however, a conical extension was added at the factory, banded by a wooden collar. *Courtesy of Mr. R. Chase.* (Value code: VR)

1-96. Most products of the Wooden Phonograph Horn Company are unmarked. This is one of the few surviving identifying seals. *Courtesy of Mr. R. Chase.*

1-97. The Sheip & Vandegrift Company of Philadelphia manufactured a quality line of wooden horns for cylinder and disc talking machines, first appearing in November 1908. These horns were marketed under the name "Music Master," and were eventually offered by Edison and Columbia as factory equipment on some of their machines. This "Music Master" horn is made of spruce, a costly option that evidently generated limited sales, even though it supposedly possessed superior reproduction qualities. This example measures 21" diameter x 26" long, and features a gold-plated metal cone section, which suggests that it once accompanied a special-order gold-plated phonograph. *Courtesy of the Gfell family collection*. (Value code: VR)

1-98. The finest product offered by Sheip & Vandegrift was the "Music Master" horn executed in mahogany with inlay and mother-of-pearl decoration. This example, measuring 21" diameter x 23" long, is mounted on a Victor "VI." (Value code: VR)

1-99. The Victor "No.31" horn (known to collectors as the "spear tip") was catalogued in oak or mahogany. Therefore, one wonders why this example was made of walnut. Perhap it was originally matched with a special-order walnut Victo from which it has been separated. It is shown here mounte on a Victor "III" in a typical oak cabinet. (Value code: VR)

Left: 1-100. Columbia, like other phonograph companies, began offering wooden horns for its machines after 1908. This 3 1/4"x 6 3/4" card described and illustrated these accessory horns. *Courtesy of Leslie Rosenfeld.*

Right: 1-101. The reverse of the card touted the qualities of the Columbia wooden horns. One wonders if the "guarantee" stated at the end of the first paragraph would have been honored if a customer attempted to duplicate the pictured demonstration! *Courtesy of Leslie Rosenfeld.*

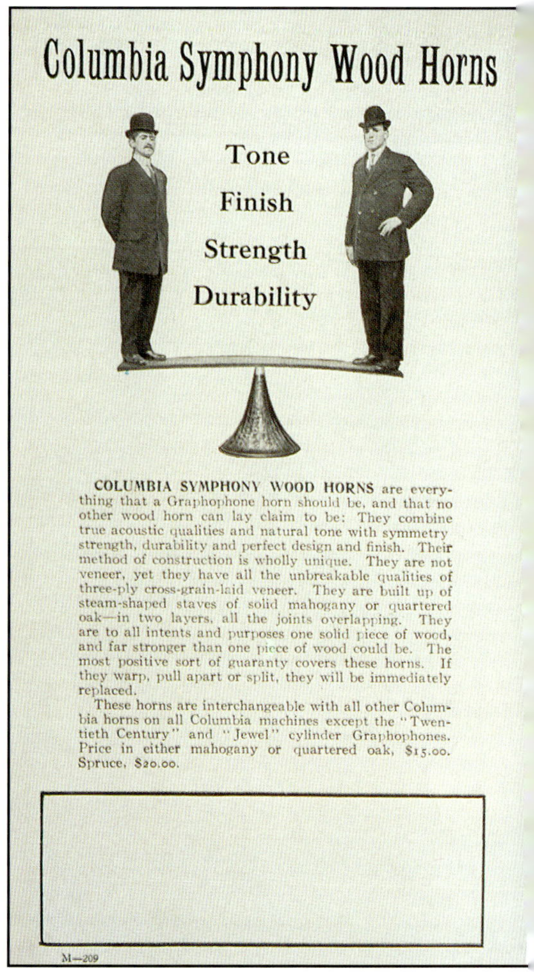

1-102. One of the least successful, shortest-lived, and most ugly horns ever marketed was the "Echo-Tone," which first appeared in the *Talking Machine World* in April 1908. Based upon a reflex principle, which would be successfully incorporated into many World War One-era machines, such as Pathé, Sonora, and British Decca, the design of the "Echo-Tone" may have been prescient, but its appearance undoubtedly discouraged sales. Advertisements for the "Echo-Tone" in the *Talking Machine World* ceased after December 1908. This example is shown on a Victor "III." (Value code: VR)

1-103. "The Resonator" was a peculiar product offered around 1910. Its purpose was to replace the horn of a Victor or Columbia talking machine, and the Talking Machine Soundboard Company of New York City claimed the device, "Reproduces Perfectly all Records. Preserving Tone Quality of Original Production." Not a terribly definitive mission statement, but refreshingly subtle. As can be seen here, the body of the attachment is made primarily of wood. (Value code: VR)

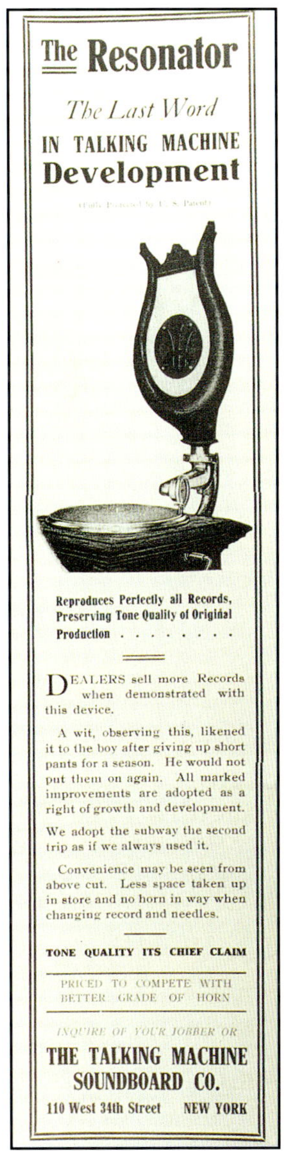

1-104. This advertisement from the September 1910 *Talking Machine World* described "The Resonator" as "The Last Word in Talking Machine Development," but it was nearly the last word from the Talking Machine Soundboard Company, too. This device, though "enthusiastically received," disappeared quickly.

1-105. This Victor "IV" features a Hawthorne & Sheble nickel-plated brass horn of the 1908-1909 period, with original rose appliqués. Although they charm us today, these are just the sort of pasted-on fancies deplored by the Tea Tray Company (see illustration 1-82). *Courtesy of Fred Guenther*. (Value code: D)

1-106. A similar Hawthorne & Sheble horn, but for a cylinder phonograph (here a Type "BK" Graphophone), and showing the "Crystal" finish offered by the firm (which in the work-a-day world goes by the unromantic term "galvanized"). *Courtesy of the Scott and Denise Corbett collection.* (Value code: C)

-107. For truly celebrating the talking machine horn, one must
cknowledge the Germans. This is an Excelsior disc machine with a
ymphonista "No.10" attachment substituted for its original horn. The
Symphonista," as the name suggested, was supposed to impart the
obustness of a band instrument to the talking machine on which it was
nounted. If it's to be believed that size matters, this contraption was a
uccess. *Courtesy of Brady Jeffcoat*. (Value code: A)

1-108. Another German machine sporting an impla sible horn attach- ment. The painted portion of the horn measures slightly wider (25") than it long (24 5/8"). The top of the upper pe is fully three feet above the turntabl *Courtesy of Richard and Jill Pope*. (Valu code: A)

1-109. This talking machine's glass horn (marked with German patents) is obscured by the cabinet, unless opened as shown. Mounted on casters, this curious creation stands 55 1/2" high. *Courtesy of Garry James*. (Value code: VR)

1-110. Closed, the cabinet's design elements bring to mind a tall case clock, with the soundbox suggesting the pendulum. Sound from the horn is reflected against the top of the cabinet. *Courtesy of Garry James*.

1-111 The glass horn, well-hidden, is encased like a jewel within the cabinet. *Courtesy of Garry James.*

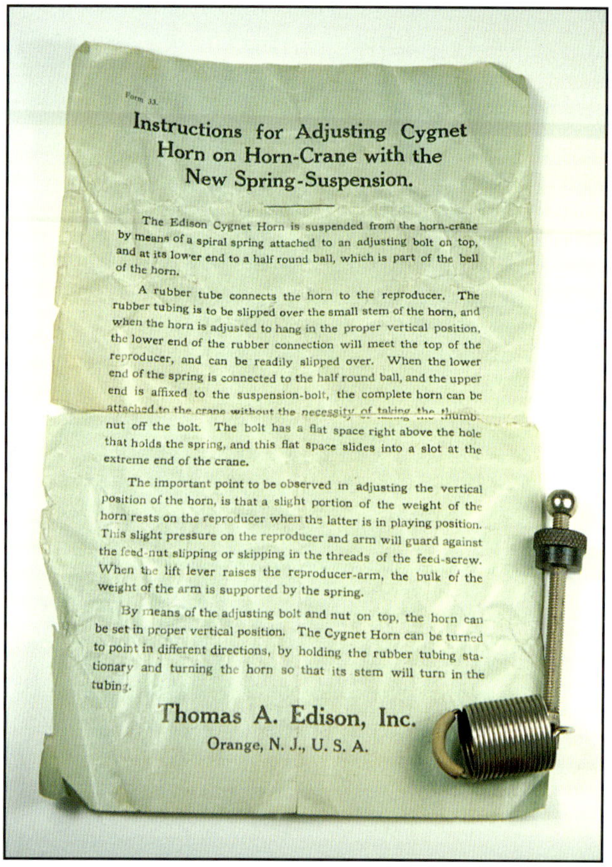

Form 33.

Instructions for Adjusting Cygnet Horn on Horn-Crane with the New Spring-Suspension.

The Edison Cygnet Horn is suspended from the horn-crane by means of a spiral spring attached to an adjusting bolt on top, and at its lower end to a half round ball, which is part of the bell of the horn.

A rubber tube connects the horn to the reproducer. The rubber tubing is to be slipped over the small stem of the horn, and when the horn is adjusted to hang in the proper vertical position, the lower end of the rubber connection will meet the top of the reproducer, and can be readily slipped over. When the lower end of the spring is connected to the half round ball, and the upper end is affixed to the suspension-bolt, the complete horn can be attached to the crane without the necessity of taking the thumb-nut off the bolt. The bolt has a flat space right above the hole that holds the spring, and this flat space slides into a slot at the extreme end of the crane.

The important point to be observed in adjusting the vertical position of the horn, is that a slight portion of the weight of the horn rests on the reproducer when the latter is in playing position. This slight pressure on the reproducer and arm will guard against the feed-nut slipping or skipping in the threads of the feed-screw. When the lift lever raises the reproducer-arm, the bulk of the weight of the arm is supported by the spring.

By means of the adjusting bolt and nut on top, the horn can be set in proper vertical position. The Cygnet Horn can be turned to point in different directions, by holding the rubber tubing stationary and turning the horn so that its stem will turn in the tubing.

Thomas A. Edison, Inc.
Orange, N. J., U. S. A.

1-112 The Edison company prided itself on the way it dotted the "i"s and crossed the "t"s. Even the simplest mechanical relationship might inspire a detailed set of instructions. When Edison added an extension spring to its cygnet horn supports circa 1910 (the earlier versions lacked the flexibility a spring could provide), the firm issued this informative flyer. Observe that the lower loop of the spring (to which the horn's ball-and-socket connector attached) is sheathed in vibration-deadening rubber. Even the best maintained Phonographs are missing this delicate detail today. *Courtesy of the Scott and Denise Corbett collection.* (Value code: F)

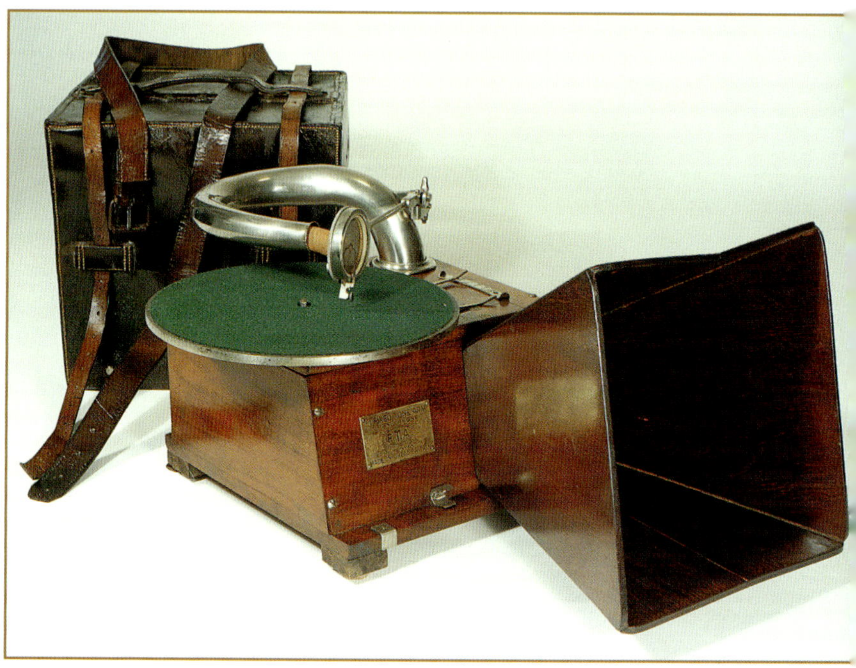

1-113. The British have a knack for making things look…well…*British*. Consider the Lancaster bomber, the London taxi, and the "Fixed Tone Arm Gramophone." This machine, with collapsible wooden horn, fit within the carrying case shown in the background. This device bears certain similarities to the Edison Bell "Primaphone" advertised by J.E. Hough, Limited, London, in the August 1909 *Talking Machine World*. The "Primaphone" employed a folding horn (not identical), and a "fixed" arm which was described as follows: "The sound-box itself is caused to advance and by an insulated Telescope connection the music, etc., is projected direct into the Taper Tone-arm and finds a clear passage without joint or obstruction of any kind into or through the…horn." The same principle is evidenced in the "FTA" machine shown here. The "insulated Telescope connection" is the rubber tube that runs between the soundbox and arm. *Courtesy of Garry James.* (Value code: A)

1-114. To quell the expected wave of usurpers attempting to take credit for this achievement, the manufacturer proudly affixed this plate to the "FTA" Gramophone. The rightful owners were prepared to defend their invention against all comers. As Americans say, "You've gotta love these guys...!" *Courtesy of Garry James.*

1-115. The Edison "Fireside" Phonograph was introduced in 1909 ($22.00) as an "up-dated" version of the Edison "Standard." Its mechanism would prove enduring – manufactured as an external horn Phonograph until the end of 1913 (after which existing stocks continued to be sold into 1915), then incorporated into the design of the lesser-priced Amberolas. After December 1912, Edison equipped some "Fireside" Phonographs with a conventional Edison horn crane as supplied with several other models. This enabled owners of the "Fireside" to mount "Blue Flowered Horns." However, an extra "horn crane base" (right) was also included as a safety measure. The December 1912 issue of the *Edison Phonograph Monthly* explained, "This crane base will counterbalance the extra weight of the horn and will minimize the possibility of tipping the machine forward." These blue horns were, in fact, the same as the conservative black-enameled "Standard" horns that had been available since October 1907. Standard Metal Manufacturing Company (the similarity of the name only coincidental), which produced the horns for Edison, was engaged to refinish them in an appealing way. (Value code: D, crane only) *Courtesy of Garry James.*

1-116. In September 1917 – the twilight of phonograph horn manufacture – a Mr. C.R. Light of San Francisco made a query to the Standard Metal Manufacturing Company on the subject. The company responded with a letter and a descriptive catalogue.

1-117. The Standard catalogue must have been printed more than a decade before, as it illustrates horns obsolescent by 1906.

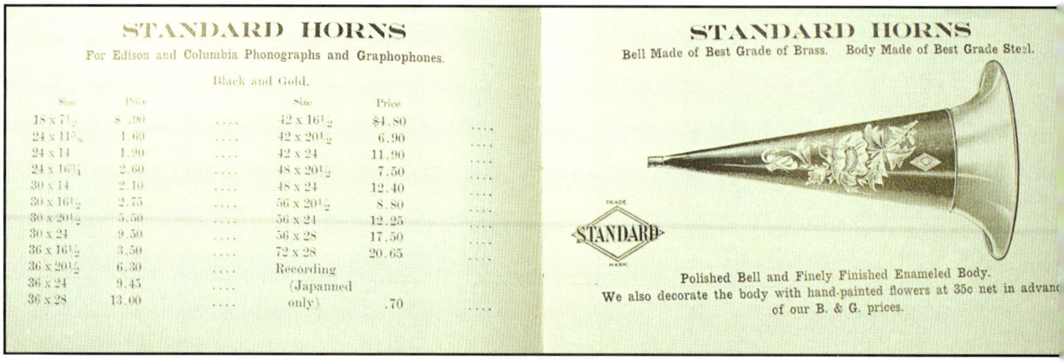

STANDARD HORNS

For Edison and Columbia Phonographs and Graphophones.

Black and Gold.

Size	Price		Size	Price	
18 x 7½	$ .90	....	42 x 16½	$4.80	....
24 x 11⅝	1.60	....	42 x 20½	6.90	....
24 x 14	1.90	....	42 x 24	11.90	....
24 x 16¾	2.60	....	48 x 20½	7.50	....
30 x 14	2.10	....	48 x 24	12.40	....
30 x 16½	2.75	....	56 x 20½	8.80	....
30 x 20½	5.50	....	56 x 24	12.25	....
30 x 24	9.50	....	56 x 28	17.50	....
36 x 16½	3.50	....	72 x 28	20.65	....
36 x 20½	6.30	....	Recording		
36 x 24	9.45	....	(Japanned		
36 x 28	13.00	....	only)	.70	....

STANDARD HORNS

Bell Made of Best Grade of Brass. Body Made of Best Grade Steel.

TRADE STANDARD MARK

Polished Bell and Finely Finished Enameled Body.
We also decorate the body with hand-painted flowers at 35c net in advanc of our B. & G. prices.

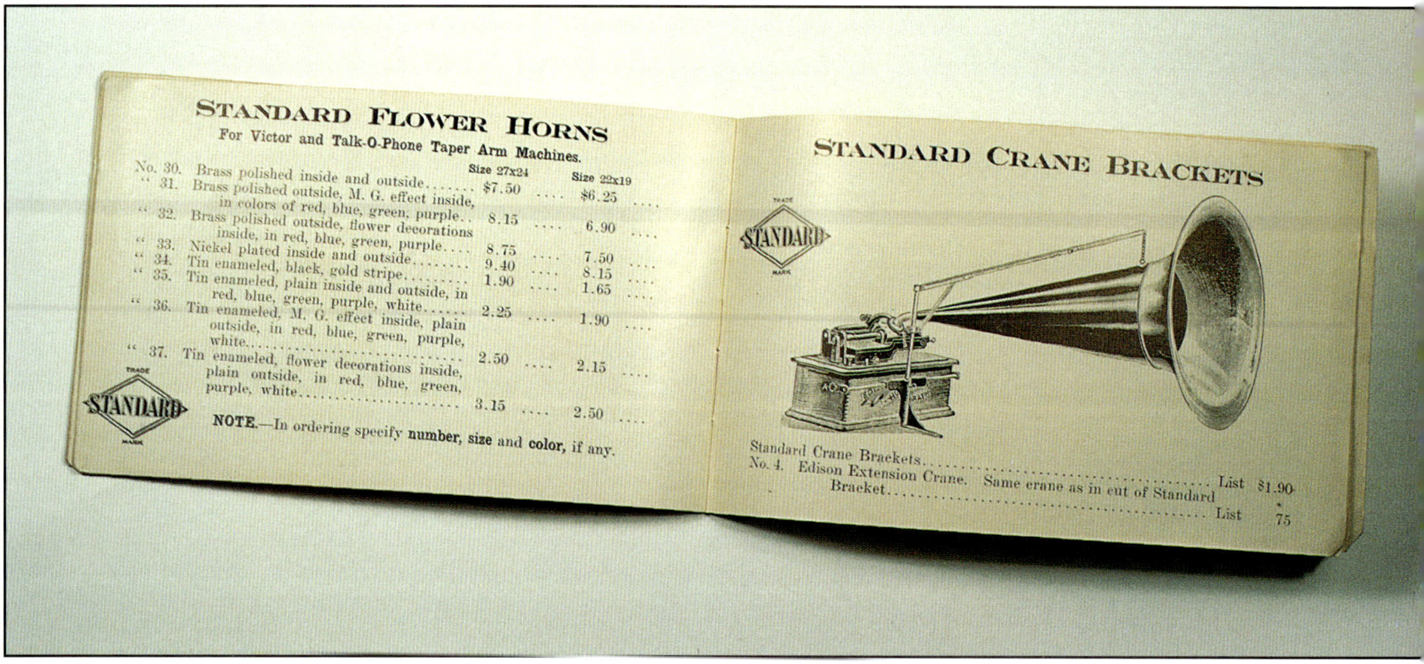

STANDARD FLOWER HORNS

For Victor and Talk-O-Phone Taper Arm Machines.

		Size 27x24		Size 22x19	
No. 30.	Brass polished inside and outside	$7.50	....	$6.25	....
" 31.	Brass polished outside, M. G. effect inside, in colors of red, blue, green, purple	8.15	....	6.90	....
" 32.	Brass polished outside, flower decorations inside, in red, blue, green, purple	8.75	....	7.50	....
" 33.	Nickel plated inside and outside	9.40	....	8.15	....
" 34.	Tin enameled, black, gold stripe	1.90	....	1.65	....
" 35.	Tin enameled, plain inside and outside, in red, blue, green, purple, white	2.25	....	1.90	....
" 36.	Tin enameled, M. G. effect inside, plain outside, in red, blue, green, purple, white	2.50	....	2.15	....
" 37.	Tin enameled, flower decorations inside, plain outside, in red, blue, green, purple, white	3.15	....	2.50	....

TRADE STANDARD MARK

NOTE.—In ordering specify **number, size** and **color,** if any.

STANDARD CRANE BRACKETS

TRADE STANDARD MARK

Standard Crane Brackets.............................List $1.90
No. 4. Edison Extension Crane. Same crane as in cut of Standard Bracket.............................List .75

1-118. Not only horns, but also phonographs of 1906 were pictured. Note the crane bracket on right page.

1-119. This “Parti View of Decoratir Room” dispels th myth that such ar was exclusively a product of the distaff side. In the background, men can be seen decorating trump style and flower horns.

VICTOR ATTACHMENT

Japanned black. This is a simple extension to use on Victor B. & G. Horns to have them fit Edison or Columbia machines. A Ring and Eye (which a dealer must attach to body of horn) goes with this attachment. This enables the trade to work off. B. & G. Victor Horns now useless.

TRADE STANDARD MARK

ALUMINUM HORNS

Highly polished or Satin finished. Especially adapted for small Talking Mashines only. Will be pleased to correspond with you for particulars if interested. We can also make various shapes and sizes of same or similar metal.

1-120. An interesting page from the Standard horn catalogue shows a "Victor Attachment" designed to convert older Victor trumpet-style horns ("Black and Gold") to cylinder horns. At the bottom is shown a small aluminum flared horn, long thought by modern collectors to be strictly European. The use of this catalogue during the First World War period suggests that Standard either maintained a diverse inventory of older products, or was willing to manufacture to order for the prices listed.

1-121. Nipper's got nothing on this charmer. *Courtesy of Walter and Carol Myers.*

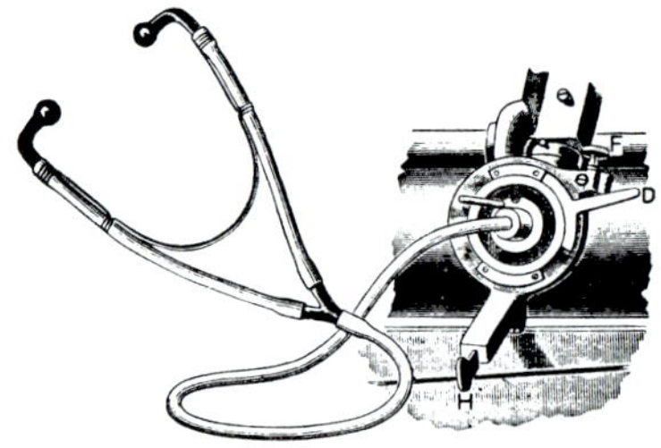

Two:
Cylinder Talking Machine Gadgets

The relative scarcity of cylinder phonograph peripherals can be understood when one considers that there was only one decade of parity between cylinder and disc products before the utter market domination of the latter. The fact that this decade (1905-1915) corresponded to the talking machine's emergence from its formative stages explains the charming, unsophisticated characteristics of related aftermarket products. The disc phonograph accessory market of the late teens and the twenties enjoyed years of refinement that the cylinder market did not. As you compare the products illustrated in this chapter with those of Chapter Three, the contrasts in design and sophistication will become evident.

It was just as disc gadgetry was really picking up steam that the declining popularity of the cylinder record discouraged further development of cylinder accessories. By the close of the teens, the disc market was booming. There remained only two American manufacturers of cylinder records: Edison and the Indestructible Record Company of Albany, New York. After 1922, Edison would continue alone. So, at the outset of an energetic decade of disc talking machine activity, 1919-1929, the cylinder phonograph was merely a shadow. We observe the trajectories of the two formats crossing in the mid-teens – one trailing downward, the other arching upward toward the era of electronic wizardry.

2-1. Primordial phonograph gadgetry: a shred of recorded tinfoil, used to promote an exhibition in Chicago. *Courtesy of René Rondeau.* (Value code: VR)

The Speaking Phonograph

OR

TALKING MACHINE

Is the Invention of

THOS. A. EDISON, of Menlo Park, N. J.

It successfully records and reproduces the human voice, thereby performing the functions of the ear and vocal organs in the human body.
It is purely mechanical in its actions, and very simple.

It Talks, Laughs, Sings, Coughs, Whistles,

Produces the sounds of Animals, and

PLAYS CORNET SOLOS.

It can be seen at the

Merchants' Building

Cor. Madison & Dearborn Sts.

And the Price charged for Admission is

TEN CENTS.

☞ Every man, woman and child should see this great curiosity.

Blakely, Brown & Marsh, Printing House, 155 Dearborn St.

2-2. The reverse of the tinfoil announcement, with exaggerated verbiage. *Courtesy of René Rondeau.* (Value code: VR)

2-3. The Edison Talking Doll of 1890 may have performed a nosedive into economic oblivion, but the doll's subsequent obscurity has made surviving artifacts all the more precious. Shown is the playing mechanism, usually missing from existing doll bodies. A simple, hand-driven affair, this assembly wore out the record after repeated playings, rendering the expensive metal and wooden doll unusable. *Courtesy of the Julien Anton collection.* (Value code: VR)

Directions for Winding.

Remove tape which binds back-spring to winding-shaft, place key on shaft and wind by turning to the right, keeping music count 1-2-3-4. The doll will talk slow or fast, and the regular time can be easily acquired by a little practice.

After it has ceased talking, pull down the baek-spring and it is ready to talk again.

Packed by

[OVER.]

2-4. Even rarer than the Edison doll mechanisms are the original instructions for playing. *Courtesy of the Tom and Sandi McCarthy collection*. (Value code: VR)

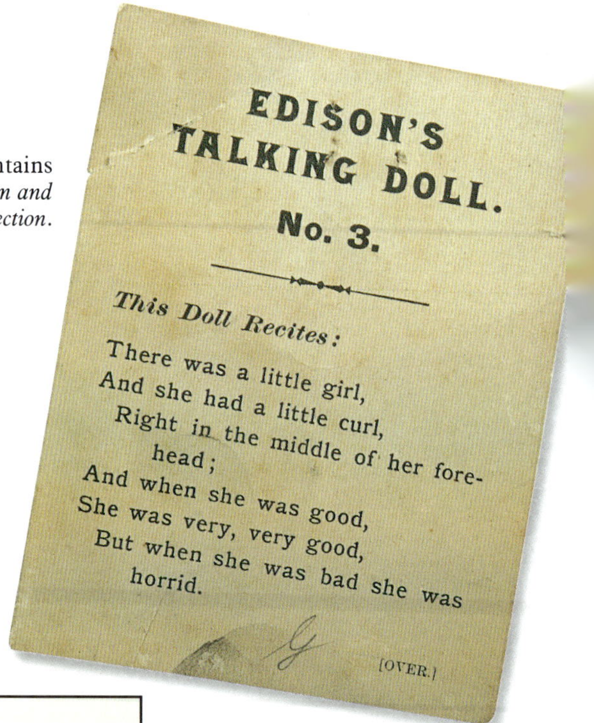

EDISON'S
TALKING DOLL.
No. 3.

This Doll Recites:

There was a little girl,
And she had a little curl,
Right in the middle of her fore-
head;
And when she was good,
She was very, very good,
But when she was bad she was
horrid.

[OVER.]

2-5. The reverse of the Edison doll instructions contains the text for the record provided. *Courtesy of the Tom and Sandi McCarthy collection.*

2-6. Edison's very first wax cylinder Phonographs intended for commercial application (through a leasing arrangement by the North American Phonograph Company) were equipped with the so-called "Spectacle" carriage. Both reproducer and recorder were carried by a rather complicated arm, of which this illustration is proof. The "Spectacle" device was discontinued after November 1889. Here, we can see an extremely early variation of the built-in shaver that until 1907 would be a key part of most Edison Phonographs. At the center of this picture, immediately below the reproducer at the left, the arching shaver arm can be seen, terminating in a narrow sapphire "knife." (Value code: VR)

2-7. It surely occurred to cylinder record manufacturers early on that a suitable method of identifying the contents of a cylinder would make everybody's life easier. Prior to the advent of cylinder mass production, which allowed the title and artist to be molded into the record rim, some companies went to considerable lengths to mark cylinders individually – Pathé in France had each record inscribed in longhand. The reduced segment near the top of this Bell-Tainter ozocerite-coated cylinder, circa 1890, suggests it was intended to hold in place a band of paper on which a notation could be made. Cylinder measures 1 5/16"x 6". *Courtesy of Gilbert "Bert" Pasley.* (Value code: VR)

2-8. During the period of the North American Phonograph Company, which had the franchise to lease (and later sell) both Phonographs and Graphophones 1889-1894, cylinder record "blanks" manufactured by (and often recorded by) Edison were generally identified by their "channel rim." Edison-made blanks were also popular with other recording firms throughout the early and mid-1890s (a time when Columbia was experiencing certain difficulties in achieving quality cylinder production). The raised edges of the Edison blank sometimes held an identifying paper ring such as the one pictured here. Ordinary use, carelessness and decades of storage robbed most records made on Edison blanks of these ephemeral labels. *Courtesy of J. Palinsky.* (Value code: VR)

2-9. The Graphophone was an abject failure in its original incarnation. It had been introduced in 1889, employing a wax-coated cardboard cylinder record measuring 1 5/16" x 6". Beginning in 1893, the Graphophone Company attempted to cheat fate by adapting some of its non-saleable inventory of machines to record and play the 2 1/8" diameter by 4" long solid wax cylinder that Edison had devised. The Edison style cylinder had proved to be the only commercially feasible recording format at that time. Since the lightweight cardboard Graphophone cylinder had not required the support of a mandrel, the principal addition to the Graphophone mechanism was the fitting of a mandrel to accommodate Edison cylinders. Mandrels were fashioned of metal or, less commonly, the cloth-covered wood shown on this Type "F." *Courtesy of Bert Gowans.* (Value code: VR)

2-10. A close-up of the wooden Graphophone mandrel shows the delicate execution of the turning. *Courtesy of Bert Gowans.*

2-11. In the early-to-mid 1890s, when the Graphophone Company adapted some of its moldering inventory of talking machines to acce the Edison-style cylinders, certain alterations we often made to the tracking mechanisms. Some Graphophones were converted to track in the Edison format (100 threads-per-inch), while othe retained their original 160tpi gearing. Many Graphophones were adapted to the Edison tpi at this time, and some were left as they were (since talking machines were used almost exclusively fo office work, if one recorded and played back usin the same machine, the tpi wouldn't matter). However, at least one Graphophone, the Type "K" shown here, was made convertible to either Graphophone or Phonograph tpi. To hold the gears, a convenient little drawer was created that slipped into the side of the cabinet. *Courtesy of Ra Phillips.* (Value code: VR)

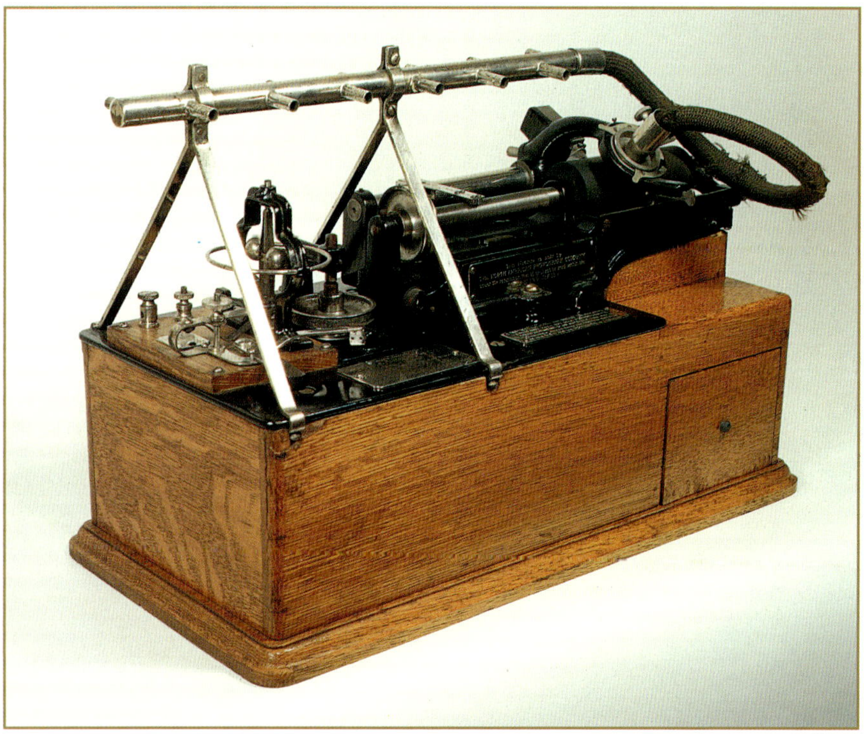

2-12. Rubber "listening tubes" concentrated the sound of a phonograph, making it easier to appreciate. After 1902, with improvements in records and reproducers, these ear tubes became largely obsolete – except in coin-slot arcades where the sound from adjacent machines would conflict. At the turn of the twentieth century, however, listening tubes were still in general use. The ingenious "A-frame" attachment shown here allowed 13 sets of ear tubes to be applied to this Edison Class "M" electric Phonograph (in later style cabinet). Electrically-driven machines such as this had been used by exhibitors since the early 1890s, with sets of tubes "rented" to patrons – usually at a nickel a record. (Value code: VR)

2-13. In the 1890s, a convenient power source for electric talking machines was the Grenet battery cell. Although use of these glass batteries in the United States had all but ceased by the turn of the twentieth century, they continued to be popular in Europe for another decade. In most cases, the glass jar was not marked or decorated. However, in this unusual instance, an etched pattern resembling foliage may be seen. (Value code: B)

2-15 (below). In 1896, this "Echophone" cylinder talking machine was shipped via Wells Fargo from W. Hill & Company of Chicago to D.H. Soule of Little Shasta, California. You see it here as it might have looked when Mr. Soule first cracked open the wooden shipping crate. Take special note of the conical horn. It resembles a typical black-japanned horn of the style sold to accompany inexpensive cylinder instruments during the late 1890s and early twentieth century. The small end of the horn, however, has a coupling of particularly narrow diameter; a non-standard fitting required for the "Echophone." The machine itself is also quite a peculiarity. The inventor, Edward H. Amet, designed certain features to discourage patent litigation by companies holding the rights to fundamental elements of cylinder talking machine construction. Not the least of the bizarre particulars of the "Echophone" is a hollow glass arm, the extremity of which is formed into a stylus that reproduces the record. *Courtesy of Gilbert "Bert" Pasley.* (Value code: VR)

-14 (above). This well-made oak carrying ase with brass hardware precisely fits the 890s Class "M" Edison electric Phono- raph seen snuggled inside it. Most people ried to avoid lugging around a 60 pound alking machine, but itinerant exhibitors ad to find means of transporting them. *ourtesy of Gilbert "Bert" Pasley.* (Value code: 'R)

2-16. The small shaving attachments which were supplied with many home entertainment cylinder talking machines prior to 1907 were notoriously insufficient to the task of shaving records. Edison offered a simple hand-driven domestic shaver unit in 1912. However, at the turn of the twentieth century, Gianni Bettini sold this more technically sophisticated manual shaving mechanism. An American catalogue listed it for $23.00 – which was considerably more than many talking machines cost at the time! Note the camel hair brush for removing the wax swarf. *Courtesy of the Julien Anton collection.* (Value code: VR)

2-17. A necessity before the advent of molded title information, the humble cylinder record title slip has here been nobly embellished by Signor Bettini, master of the hoity-toity. *Courtesy of the Scott and Denise Corbett collection.* (Value code: VR)

2-18. It's curious that cylinder record manufacturers of the 1890s didn't adopt the simple method of identification used by this obscure New England brand. Prior to the introduction of the molding process 1901, no information, not title nor artist nor catalogue number nor matrix, was pressed into the surface of the record. The inclusion of a spoken company name supposedly guarded against the possibility of fraud (though it should be noted that during the mid-1890s Columbia sold records which omitted a company name from the announcement). Ostensibly, the spoken announcement was begun on cylinders because of the necessity for positive identification. However, a closer look at the spoken announcement shows it to have been not only a protection of intellectual property, but also a social superimposition. The very idea of an articulated introduction to proceed a musical selection is the quaint product of the nineteenth century mind. Polite, stodgy, and rhetorical, the announcement gave the phonograph a pious note of respectability. Like disguising the automobile as a buggy, the announcement was a typically Victorian reaction to science, that gave the otherwise rude and unnerving chatterbox a semblance of grace and propriety. *Courtesy of Michael Sims.* (Value code: F)

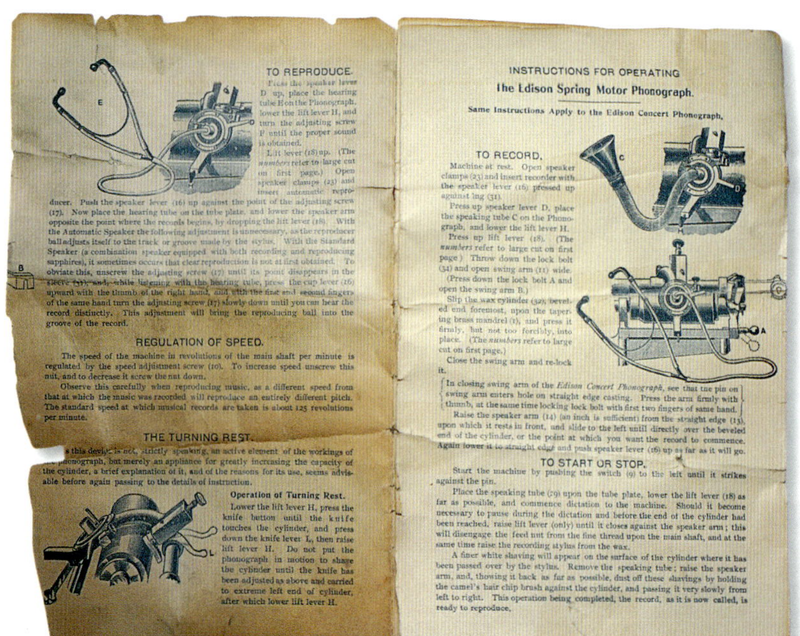

2-19. There were a number of different listening tube configurations. Most common were the simple rubber ear-tubes reinforced internally by wire. However, as these instructions for an Edison "Spring-Motor" Phonograph show, some listening tubes were held in position by an external "bridge" (see illustration 2-97 of *The Talking Machine, an Illustrated Compendium*). (Value code: G)

2-20. These delightful lithographed cards were among the various subjects used as premiums for turn of the twentieth century European products such as Liebig meat extract. The public was encouraged to collect entire series ("famous inventors," etc.) by frequent patronage of the product. Of particular interest is the card at the lower right, which depicts an artist's impression of acoustic recording in its early stage of development. Banks of phonographs simultaneously recorded a performance, a process that had to be repeated in order to obtain a cache of first-generation cylinders from which multiple copies could be duplicated. Also note the Edison Phonograph laboratory as it was imagined at the upper right. The quaint technology amid a neat, formal domestic setting appears to be inspired more by the works of H.G. Wells or Jules Verne than by grimy reality. *Courtesy of Yves Rouchaleau.* (Value code: E)

2-21. This attachment automatically stopped the motor of an Edison "Home" Phonograph Model "A" when the reproducer carriage reached a predetermined point. The device was fixed in place by using one of the bolts which secured the honograph's upper casting. Spring tension as the means by which the brake lever was pulled into the "off" position (as shown). (Value code: D)

2-22. Two views of the same type of cylinder record box. These labels were locally printed by a dealer and pasted over the first incarnation of the "Miss Columbia" box label, circa 1897. Illustrated is a Columbia Type "A" Graphophone, a machine that made a significant contribution to the cylinder talking machine industry of the late 1890s by bringing the price within reach of the middle class. *Courtesy of George Glastris.*

2-23. It is utterly astounding how much inventive genius must have been employed at the Excelsior phonograph works in Germany. The number of different cylinder models that were designed and produced there, each elevated by some imaginative touch, seems beyond documentation. Every now and again yet another fantastic mechanical creation from this factory is discovered to exist. Shown here is an Excelsior, circa 1900, the base of which appears to have been inspired by an ordinary Type "A" Columbia Graphophone – a machine that Excelsior commenced in business by imitating. However, the "egg-beater" mandrel, reproducer carriage arm and sewing machine style flywheel are all unquestionably distinctive. The listening rail, with nipples for attaching multiple ear-tubes is also highly original. To top it off, a spirit level may be seen at the rear of the upper casting, just to the left of the flywheel. *Courtesy of Jalal and Charlotte Aro.* (Value code: A)

2-24. The physical link that communicated the sound vibrations fro the stylus to the diaphragm of an acoustic talking machine reproduc was the focus of considerable imagination. The Graphophone Company, from about 1893 till 1904, attached the stylus to a lateral arm extending across a radius from the center of the diaphragm to the periphery of the reproducer frame. Ingenious minds soon attempted improve on this design by fixing a "cup" or "dome" surmounted by a stylus directly to the diaphragm. For reproducers of the Graphophor or "floating" type, this dome arrangement became most common in Great Britain and Europe. In the United States, Hawthorne and Sheble, manufacturers of talking machine accessories, encouraged clients to adapt Graphophone reproducers by removing the factory-installed stylus arm and substituting "The 'H&S' Aluminum Disc" ("Consists of an Aluminum Cone having a point set in the centre. When placed on a diaphragm of a Graphophone Reproducer it increases the volume enormously, and liquifies the tone, making it clear, loud and brilliant.") By far the most bizarre of conversions, however, was the "Ruby Point for Graphophone Reproducers." Thou unfortunately priapic in appearance, the Ruby Point was designed to precisely replicate the "offset" position that the stylus usually occupi on the Graphophone lateral arm. The pin seen to the right of the diaphragm housing originally secured the terminus of this arm. Note also the rubber stamped patent information (September 25, 1900). Th Penn Phonograph Company, 19 South Ninth Street, Philadelphia, advertised the Ruby Point in the June 1900 issue of the trade magazi *Phonoscope*. "Sells Like Hot Cakes, Only $1.00… Increases the volum 50 per cent… Consists of a glass base with hard composition tip… BEWARE of all metal imitations, with point set in, they lack the brilliancy in reproduction which so distinguishes the Ruby Point." United States patent No.658,571 was granted to Jay W. Moyer for this device. The Graphophone Company eventually accepted the wisdom of eliminating the stylus arm, and from 1904, began equipping its "floating" reproduc with directly-mounted styli. (Value code: D)

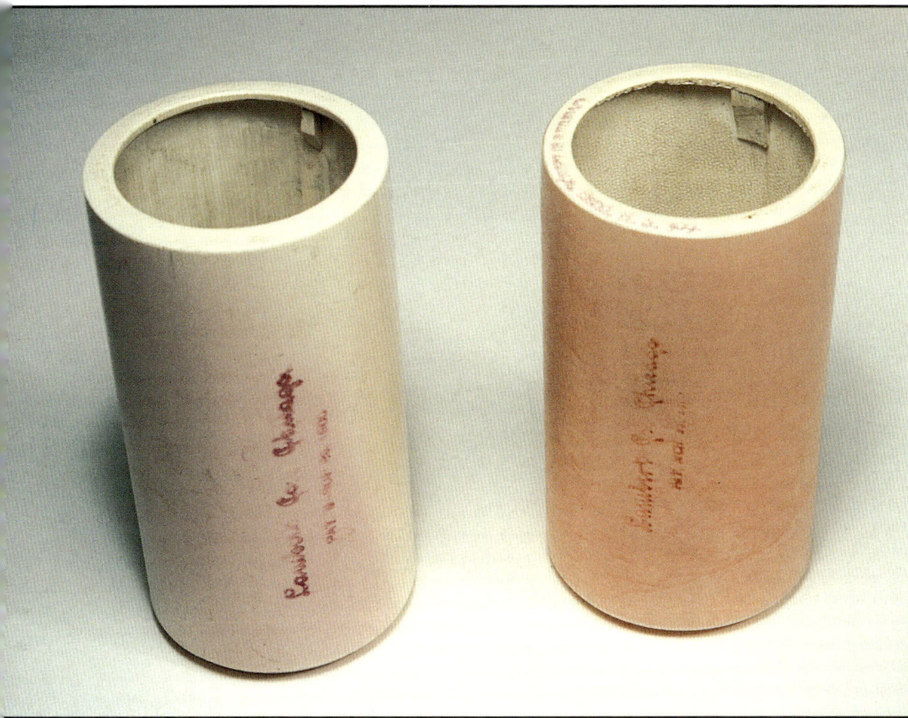

2-25. The Lambert Company of Chicago began producing durable celluloid cylinder records in 1900, at a time when every other manufacturer (except Lioret in France) was using fragile metallic soap (or "wax"). Nothing could have been more distinctive than the pink color Lambert chose to dye its records. However, to reinforce recognition of the brand, the company rubber stamped many of its early releases across the grooves, as seen here. Repeated playing or cleaning would have removed the ink, so "signed" Lambert cylinders are very rare today. *Courtesy of Michael F. Khanchalian.* (Value code: D)

2-26. Art imitating life – "Pinkie." (Value code: E) *Courtesy of Michael F. Khanchalian.*

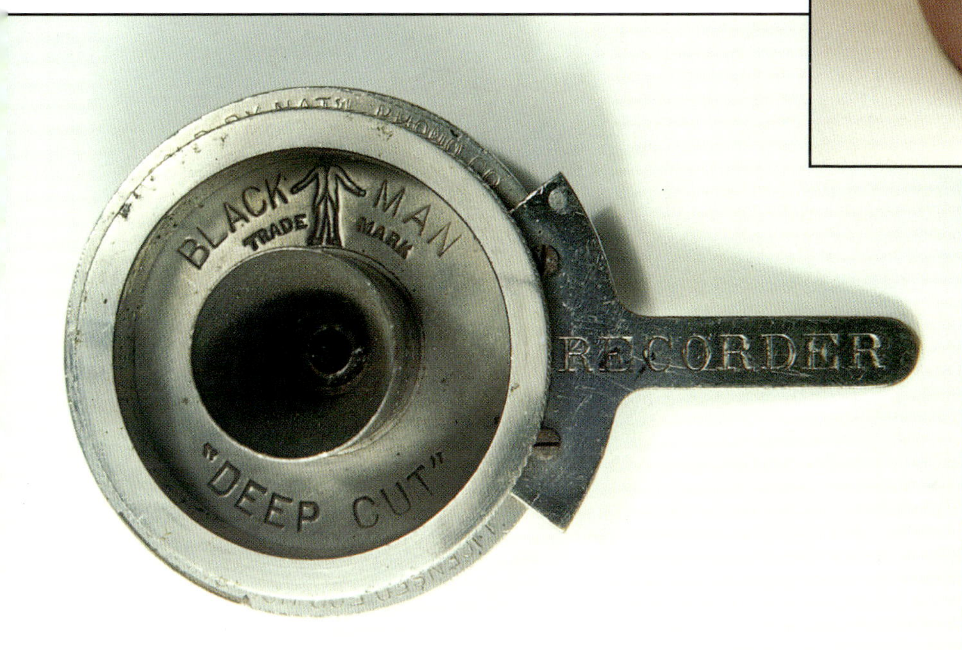

2-27. The Blackman Talking Machine Company of New York City, among its other wares, sold modified Edison "Automatic" reproducers and recorders. A dome-shaped tube plate was substituted for the factory version, which was flat and presumably less conducive to the flow of sound. This Blackman-modified Edison recorder is marked "Deep Cut." The Edison Company looked dimly upon alterations to its designs, and reproducer conversions were short-lived. Some of the marketing employed by J. Newcomb Blackman, proprietor, could hardly be considered politically correct. For example, "Blackman is White and Will Treat You Right." *Courtesy of Terry Paul Baer.* (Value code: C)

2-28. Like the Victor "experimental" soundboxes pictured in illustrations 3-55 through 3-63, these Edison reproducers were steps in the development of commercial products. On the left, Model "F" (marked with serial number 339) is virtually identical to the conventional Model "B," except for a subtle difference in the weight. On the right, Model "E" (bearing number 028) employs an early Model "B" top, attached to a perforated weight that looks like it was constructed from bits that happened to be laying around the laboratory (including an O-ring that was normally used to retain the diaphragm of an Edison reproducer). *Courtesy of the Charles Hummel collections.* (Value code: VR)

2-29. The king of bizarre cylinder talking machine attachments must surely have been the Hawthorne & Sheble "Dupliphone." This apparatus was sold to convert certain ordinary Graphophones (such as the "AT" shown here) or Phonographs (such as the Edison "Home") to accept the large, 5" diameter "Concert" cylinder records. A supplemental casting supporting a "Concert" mandrel was secured in place behind the conventional mandrel of the instrument. Power was transferred from the motor by a specially sized leather belt. A long extension tube raised the height of the reproducer to play the larger records, or the machine could be operated "normally" – hence the coined name of the device. *Courtesy of Douglas Defeis.* (Value code: VR)

2-30. To facilitate tracking of the greatly extended reproducer of the "Dupliphone," a custom-designed weight could be clipped in place (seen just above the knurled collar of the reproducer). Note also the cast information on the periphery of the 5" mandrel. An H&S advertisement in the May 1900 issue of *Phonoscope* claimed the "Graphophone Dupliphone" could be "readily applied… whereby the large Concert or Grand records can be used on the small machines… Price $16.00." *Courtesy of Douglas Defeis.*

2-31. Evidence suggests that Signor Gianni Bettini was a man of no small ego. He unquestionably considered his "Micro" attachments to be superior to the factory equipment of Phonographs and Graphophones. It is unusual, therefore, to see Bettini devices boxed with the Edison equipment it was intended to replace. Shown here is a plush container holding a Bettini Type "N" reproducer, Type "R" recorder, as well as the original Edison Model "D" reproducer and corresponding recorder from a circa-1902 Edison Concert Phonograph. Although Bettini technology included an adjustable counterbalance to vary the pressure of the stylus on the record, a supplemental weight (shown removed from its niche at the right) was provided to exert even more force. In mechanical terms, this would translate into augmented volume (though at the expense of record wear). It is fascinating to contemplate how a "Concert" cylinder, already capable of respectable volume, would sound reproduced by the Bettini "Micro-Phonograph" jumbo diaphragm with maximum weight in position. Two adjusting wrenches are seen at the sides of the box. *Courtesy of Jalal and Charlotte Aro.* (Value code: A)

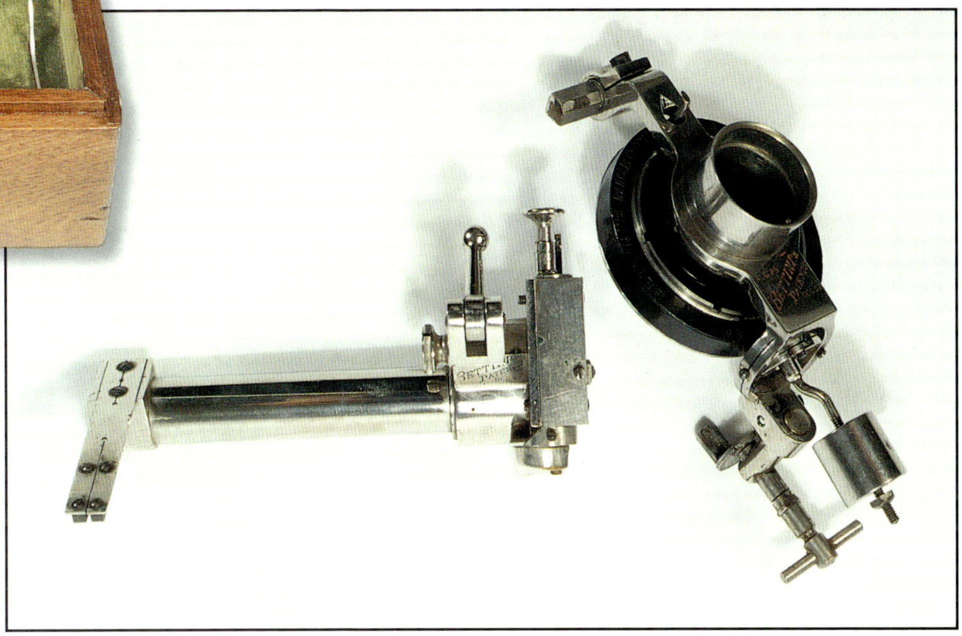

2-32. A Bettini "Micro-Phonograph" attachment consisted of a reproducer (right) and a carriage (left). The two were locked together by tightening a thumbscrew. In the instance of this Bettini apparatus for an Edison "Spring-Motor," however, a "quick-release" clutch arm has been installed, making it especially easy to detach the reproducer. Note also the conventional Edison shaving device that has been mounted on the Bettini carriage. *Courtesy of Jalal and Charlotte Aro.* (Value code: A)

2-33. From its introduction in 1899, Edison's "Gem" Phonograph was plagued by inexpensive and poorly designed reproducers – until the reliable Model "C' reproducer begrudgingly was made available for the "Gem" in 1902. During the early years of the twentieth century, however, various firms attempted to ameliorate the performance of "Gem" reproducers, principally by screwing an extra layer of metal to the counterweight. The added mass resulted in better tracking and greater volume. The thickening of this counterweight has left just enough clearance for the stylus to protrude. To the left is the corresponding "Gem" recorder, both from 1901. *Courtesy of George Glastris.* (Value code: D)

2-34. The Chicago Talking Machine Company marketed an unusual attachme called the Polyphone to fit Edison Phonc graphs and Columbia Graphophones. Th concept was simple: two reproducers tracking in the same groove a micro-seco apart would produce twice the volume of sound (through twin horns), with the onl drawback being a minor echo. In 1899 an 1900, the Polyphone was promoted, durin which time the company attempted to improve the device and make it reliable. I has been assumed that the Polyphone did not survive far into the twentieth century and that the improved, louder molded cylinders and better reproducers of the pc 1901 era certainly would have rendered th Polyphone unnecessary if it had not lost t favor of the public already. However, this Edison "Home" Phonograph and a few others like it are evidence of an unexpecte rebirth of the Polyphone. *Courtesy of the C family collection.* (Value code: VR)

2-35. An Edison "Home," late 1901, with a Polyphone attachment? Unheard of! Furthermore, observe carefully the shape and configuration of the reproducer weights. All Polyphone attachments (save this) employed conventional factory reproducers, usually with some kind of linkage appended in an attempt to keep the styli aligned. Here we see two conjoined weights that are part of the Polyphone arm itself. In fact, the arm is one with the reproducers – they cannot be withdrawn. Most telling of all, however, is the manner in which the carriage is stamped around the periphery of the reproducer housings: National Phonograph Company, Orange, New Jersey. It is known that Edison, who looked with extreme disfavor upon any outside efforts to "modify" his Phonographs, manufactured the Polyphone in 1900. Previously unknown was his intention to issue an "authorized" version of the device. The metal diaphragms were subsequently installed by a modern owner. *Courtesy of the Gfell family collection.* (Value code:)

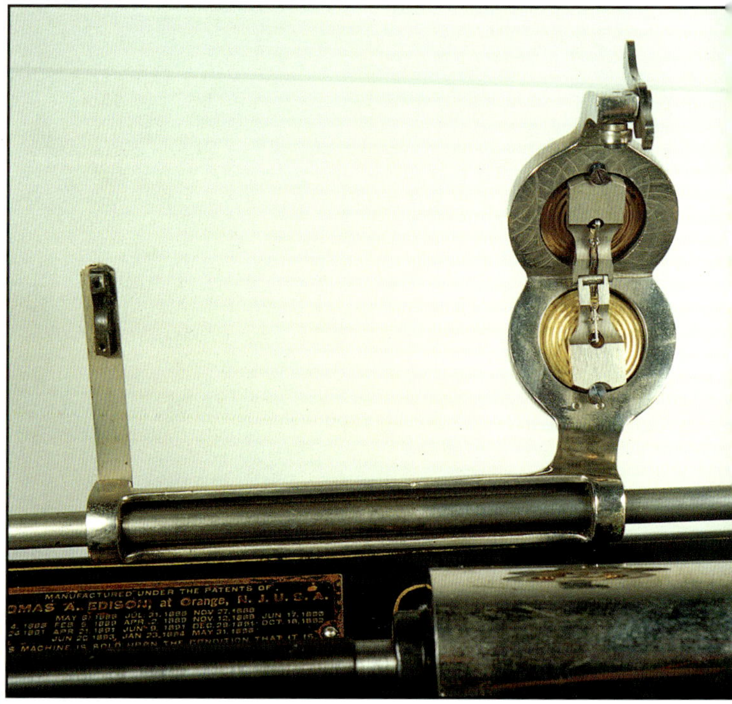

2-36. The Edison "Automatic" reproducer, marketed between 1896 and 1901, had a few unfortunate design flaws. Inventors and tinkerers soon began adapting the factory equipment to include modifications they claimed would improve the performance of the "Automatic." The most common alteration was to the counterweight, to make it heavier in order to produce greater volume. Edison reproducers had a tendency to skip or ride between the grooves because Edison was trying not to tread on Graphophone patents, which solidly protected the concept of freely articulated reproducer movement. Although the design of the "Automatic" surpassed its predecessor, the "Standard Speaker," in lateral movement, inventors were inspired to make further improvements. This adaptation of an "Automatic" by Elmer Fletcher reflects both an attempt to free the weight from constraint (note the beveled area around the limit screw in the 10 o'clock position), and to reposition the contact point of the stylus (note the oddly-shaped bar, stamped in tiny letters with the legend "Pat. Applied For." *Courtesy of Mr. R. Chase.* (Value code: C)

2-37. The domed tube plate of "The Fletcher" reproducer (an improvement over Edison's flat design) proclaims its provenance as well as the patent application that was never to be. Fletcher began adapting Edison reproducers in 1900. The May 1900 issue of *Phonoscope* reported that Douglas and Company of New York City had secured exclusive rights to market them. Edison sued, claiming that Fletcher was tampering with his product. Litigation continued for a couple of years, hampering and finally terminating Fletcher's reproducer business.

2-38. In the United States, Leeds and Catlin cylinders are among the rarest. Although the Leeds family name was associated with both cylinder and disc manufacture over a long period of time, this particular box and title slip date from around 1900. You will note the repeated use of the term "master." At the turn of the twentieth century, a marketing war of words was raging over the practice of duplicating cylinder records by direct mechanical or aural means. Cylinder duplicating had been a fact of life since the earliest days of the commercial cylinder record, in 1889 and 1890. Until 1901, there was no method in general use of molding, or mass-producing, cylinders. Therefore, primary recordings of performances were used to "dub" multiple copies. The North American Phonograph Company maintained classes of Edison records, with the more expensive presumably closer to the original "take." The independent cylinder manufacturers, such as Leeds and Catlin, Reed and Dawson, or Norcross, asserted that the only way Edison and Columbia could satisfy their big sales figures was to rely on umpteenth-generation duplicates. The semantic battle reached its zenith in 1900, with the independents advertising that every one of their cylinders was a "master," implying (or flatly stating as seen here) that they sold no duplicates. Douglas & Company of New York City advertised a "New Line of Master Records" in the May 1900 edition of *Phonoscope*. The same issue saw Norcross claiming "Makers of High Standard Master Records." The previous month, the Concert Phonograph Record Company of New York had advertised its "Autocrat Concert Record" with a cartoon showing a swank dude driving a talking machine automobile labeled "ORIGINAL" overtaking a country bumpkin operating a conveyance impugned by the word "DUPLICATE." (An autocrat is not a terribly sympathetic figure, but the trade name "Aristocrat" must have been taken.) The exact definition of what constituted "originality," and the logistics of maintaining a sufficient inventory of "master" records naturally was not explored in the advertising. With the introduction of molded cylinders, the verbiage of the independents became very much beside the point, as, unfortunately, did they. *Courtesy of the Scott and Denise Corbett collection.* (Value code: VR)

2-39. Biglow and Maine records, thc they were produced only briefly dur 1899, have a rather complex background. Hymnbook publishers Bigl and Maine sold cylinders of sacred singer Ira D. Sankey. Sankey was internationally known, and some ch hymnbooks in parts of Great Britain still known as "Sankeys." The recor ings were taken by Leeds and Catlin and duplicated at Edison's record facility (the irony of which can be appreciated by reviewing the previo caption). Most noteworthy within th context of this book, however, is the manner in which the cylinder at the center is wrapped. At the time, cardboard record containers were not lin with any material to protect the reco Instead, the record itself was rolled i sheet of cotton batting, and then frequently enveloped in tissue paper. Even the most fastidious talking machine owner could have had troub keeping his records neat – so the virginal quality of the item shown he is truly remarkable. *Courtesy of Micha F. Khanchalian.* (Value code: VR)

2-40. A Biglow and Maine brochure featuring the distinguished countenance of Ira D. Sankey. *Courtesy of Michael F. Khanchalian.* (Value code: VR)

2-41. This exquisite velvet-lined carrying case is indicative of how much talking machines were revered as scientific instruments in Europe. To the casual observer, these reproducing and recording heads, with various accessories, might appear to be Bettini products. They are, however, beautifully rendered imitations of the period. *Courtesy of the Julien Anton collection.* (Value code: VR)

2-42. This automatic stop, with its delicate superstructure, was specifically designed to fit an Edison "Home" Phonograph Model "A," such as the 1901-style shown here. The device is fixed between the upper casting and motor-mounting plate of the machine. When the reproducer carriage nudges the arm at the right (which can be adjusted to the length of the recording), a catch is released allowing an extension spring to pull the brake lever into the "off" position. This clever contraption is easily installed or removed from the instrument. (Value code: D)

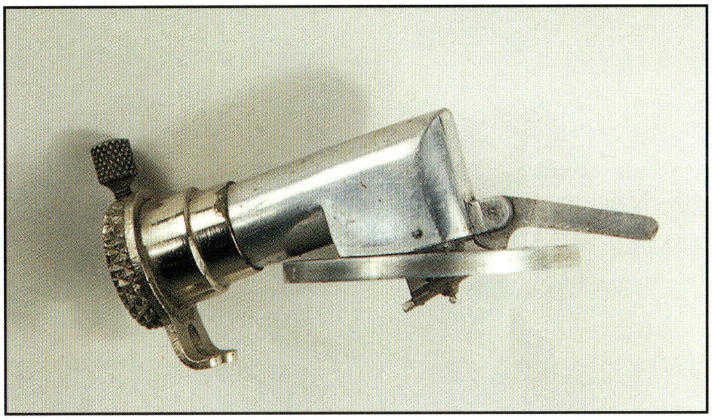

2-43. Here is early evidence of Columbia's clever efforts at "dual purpose" technology. This "floating" reproducer/recorder, which certainly dates from the 1902 period, employs a lever to cock the styli into their positions. The shallow depth of the reproducer housing associates this apparatus with office dictating equipment. *Courtesy of the Babcock House Museum, John H. Perschbacher.* (Value code: D)

2-44. The provenance of this truly bizarre artifact we do not know. "Brown" (actually a variety of colors from tan to dark chocolate) wax cylinders were the standard of the industry until the introduction of molded records in 1901. Although Columbia briefly continued manufacturing cylinders in brown wax after the molding process was adopted, Edison switched immediately to black. Then why is this Edison molded record of such a pale complexion? *Courtesy of the Babcock House Museum, John H. Perschbacher.* (Value code: VR)

2-45. During the teens, Pathé exploited the slogan "No Needles to Change" to emphasize the value of its vertical cut record system, which employed a permanent sapphire stylus. Ten years earlier, however, a similar phrase appeared on this label affixed to the top of a pasteboard cylinder record storage box. *Courtesy of the Scott and Denise Corbett collection.*

2-46. Many years before it would achieve a darker reputation as a seat of war, the Vietnamese city of Hanoi was a cultural center in South East Asia. At this exposition of early twentieth century wonders, the French Pathé company was well-represented. Every sort of cylinder talking machine can be seen, from the diminutive "Le Gaulois" with spun aluminum horn, to the massive "Le Céleste" (behind which the man on the right is standing), sporting a fluted glass horn. (Value code: G)

2-47. An assortment of unusual cylinder record boxes. Left to right: from Britain, the "Ebonoid" (with a Pre-Raphaelite flavor), the International Phonograph and Indestructible Record Company, Limited (which were all illegally dubbed from Edison records), the "Markona" (in octagonal box); from Spain, the Hugens y Acosta (with brightly lithographed metal container); and finally a cylinder of non-standard length featuring the voice of Pope Leo XIII. This recording of a portion of the Latin mass was recorded on February 5, 1903 by Gianni Bettini, and marketed by Columbia in a special white box. *Courtesy of Ray Phillips.* (Value code: VR)

2-48. Three-quarters of this cylinder record box is devoted to a map for finding the Northampton Sales & Exchange (England) – a good thing to have in a very old city. *Courtesy of George Glastris.* (Value code: G)

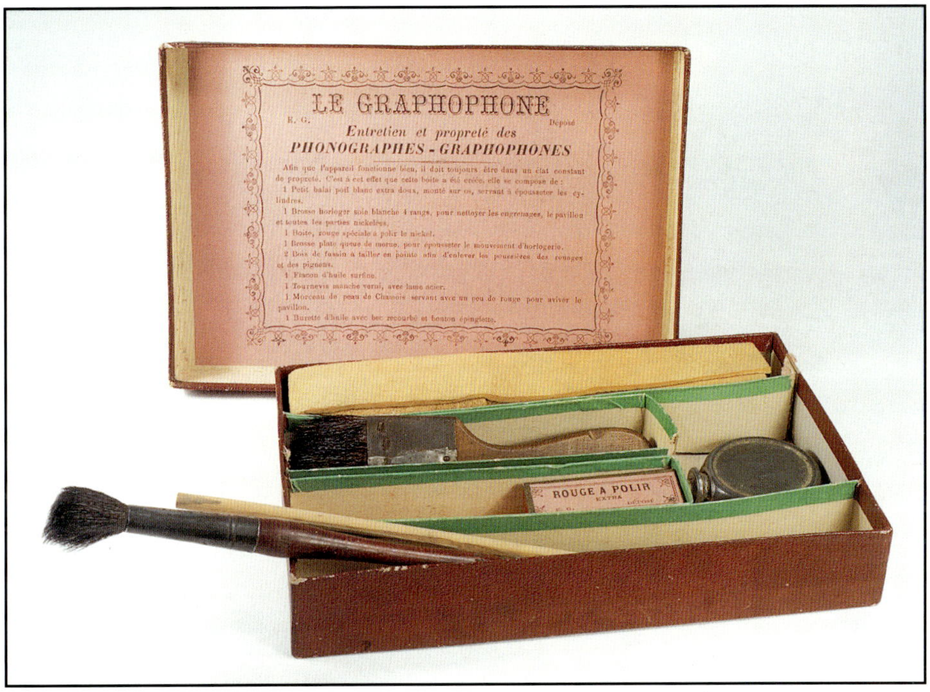

2-49. The Europeans, especially the French, had delicate sensibilities when it came to details. This charming pasteboard box contained everything the talking machine owner needed, including: oiler, polishing compound, chamois, and various brushes. Truly "une boite à nécessaire" – an indispensable accessory kit. *Courtesy of the Julien Anton collection.* (Value code: E)

2-50. A Bettini reproducer and recorder in an elegant box, typical of his Paris years. *Courtesy of the Julien Anton collection.* (Value code: VR)

2-51. As Gianni Bettini struggled to survive in Paris during the first decade of the twentieth century, he started to move into product areas that, a few years prior, would have been unthinkable. Bettini began to emblazon his name on cylinder reproducing and recording apparatus far more ordinary than his expensive "Micro-Attachments." At first, his conventionally designed merchandise retained the distinctive character of his earlier work; later only his famous name distinguished the items he sold from other anonymous European equipment. While he still clung to his standards, Bettini marketed this amazing "floating" reproducer, fashioned of beautifully finished wood, with a cut-down version of the "spider" stylus holder that made his reputation. The diaphragm bears his signature. At the right is a Bettini recorder, for use on any Graphophone-type machine (but most specifically the Bettini "No.8"), molded of green phenolic material. *Courtesy of the Julien Anton collection.* (Value code: VR)

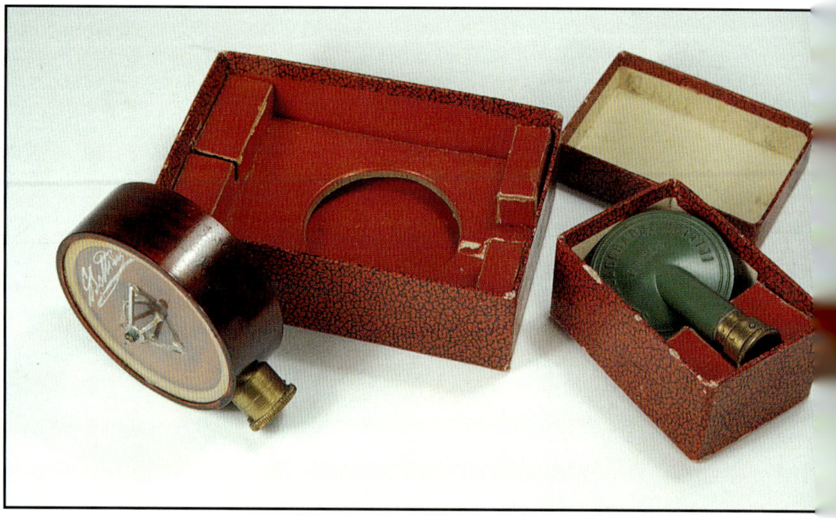

2-52. Two German lithographed cylinder record tins, "Triumph" and "Colonial." The lid of one has been removed to show how the record has been wrapped in cotton batting to protect it. *Courtesy of the Charles Hummel collections.* (Value code: VR)

2-53. A sheer delight, but fragile as glass, is this pasteboard carton to house a French "Styx" phonograph. The exterior is covered in wood-grained paper; the interior is charmingly self-promotional in art nouveau. Needless to say, the horn had to be transported separately. In the United States, Columbia offered the humble Type "Q" Graphophone in a pasteboard carrier, but without the aesthetic flair seen here. *Courtesy of Philippe Le Ray.* (Value code: VR)

2-54. A small French cylinder record carrier whose vivid compartments hold cylinder boxes in uncommon shades. (Value code: D) *Courtesy of Yves Rouchaleau.*

2-55. A Pathé "Coq" phonograph is shown here in its carrying case, fixed to a plush velour interior, and surrounded by its conical aluminum horn, ear tubes and containers holding the reproducer and recorder. Below, a drawer stores cylinder records. The logo of a talking machine firm might sometimes be found emblazoned on the lining of the lid. *Courtesy of Philippe Le Ray.* (Value code: A)

2-56. The cylinder phonograph is raised in triumph. This 17 1/2" female figure also holds a small object in her right hand – a reproducer? An identical statue was sold as "Le Dernier Cri" (the latest fad), one of a series of decorative French domestic accessories that celebrated landmarks in science and invention. The figures were cast of a "white metal" alloy, and then given an antique bronze finish. The statue is signed "Rancoulet." *Courtesy of Jalal and Charlotte Aro.* (Value code: VR)

2-57. "Skylark" was an inexpensive line of German-made cylinder talking machines, sold in Great Britain during the first decade of the twentieth century. The one shown here appears to be nothing more than a slightly more robust "Puck." *Courtesy of Bert Gowans.* (Value code: A)

2-58. It's unusual to find a "Puck"-inspired machine with a bent-wood cover and banner decal, but the importance of this particular "Skylark" is not connected to the cabinetry. *Courtesy of Bert Gowans.*

2-59. The significance of this instrument lies in its "feed screw." "Pucks," as well as certain other cylinder talking machines such as the Pathé "0" or the Columbia "AP," did not have mechanically-assisted tracking mechanisms. The stylus of the reproducer was carried along exclusively by the record groove. This "Skylark," however, employs a coarse spiral screw (center) to move the horn and reproducer in the direction of play. At the end of the record, the knurled thumbscrew seen just to the right of the lower governor weight is loosened, and the tracking arm can be returned to its starting point. *Courtesy of Bert Gowans.*

2-60. "New Century" talking machines were made in Europe, sold in Great Britain. A 1903-1904 catalogue of the London firm Waterfield, Clifford & Company, Limited featured a variety of German and French manufactured cylinder instruments under the name "New Century." Included were the ubiquitous "Puck," small models from the German Excelsior Works, as well as copies of the Columbia "Eagle" from Pathé. Seen here is a "New Century" "Puck" in its carrying box, with compartments to hold the machine, the records, and even the reproducer. *Courtesy of Philippe Le Ray.* (Value code: A)

2-61. The French mail-order firm of Girard adapted the uncomplicated Pathé "Coquet" phonograph by appending to it an elaborate contraption known as "Système Vérité." The results looked as if they had sprung from the pen of British cartoonist Heath Robinson (who, like Rube Goldberg in the United States, imagined fantastical clap-trap). The long horn, with its characteristic Chinese red painted body and aluminum bell tracking within a complicated framework, threatened to overwhelm the modest talking machine. In this instance, the "Coquet" was given stature by a matching wooden base with drawer to hold records or accessories. *Courtesy of the Julien Anton collection.* (Value code: A)

2-62. A European copy of Columbia's popular "Eagle" cylinder Graphophone, manufactured by Ullman, adapted with the customary wire superstructure of "Système Vérité." However, a "Combret" horn and reproducer have been added to the mechanism. The horn resembles the kind of aluminum-belled, pastel-bodied appliance most commonly associated with "Système Vérité." Notice, however, the fat elbow connecting the large-diameter reproducer. *Courtesy of Jean-Paul Agnard.* (Value code: A)

2-63. In France, Pathé developed a 3 1/2" diameter cylinder record called "Intermediate" (or "Salon" in Great Britain). Many Pathé cylinder instruments from the 1903-1908 period were equipped with "Inter" mandrels that telescoped over the standard mandrel. Thus, both types of record could be played. The necessity of changing the height of the reproducer to accommodate the two sizes of cylinder gave rise to a flood of contraptions. Pathé eventually eliminated the issue by creating a built-in "floating horn" arrangement, based on a previous device known as "Système Vérité." Yet, even the simplicity and ease of operation of the improved Pathé design did not stop mad geniuses from trying to disable and complicate it, as seen here. This Danish invention (marked "D.F.M. 64017.220") converted a new style "floating horn" Pathé to an old style "fixed horn" machine. A slide allowed the reproducer carriage to be raised in order to switch from standard to "Inter" cylinders. The extension arm assembly to the right, which normally conveyed the horn and reproducer across the record, was now rendered functionless. This paradigm was frequently evidenced in the history of recorded sound: if a company altered a design, there was always someone waiting to improve the mechanism by returning it to what it had been. *Courtesy of Jalal and Charlotte Aro.* (Value code: VR)

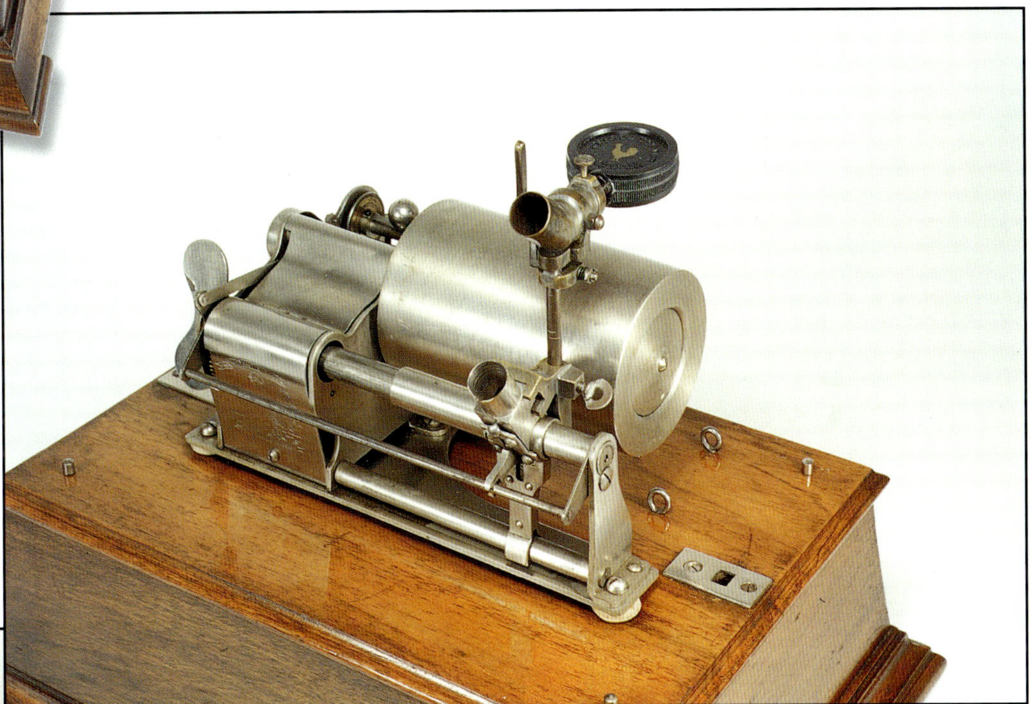

54. The Pathé company's promotion of its 3 1/2" diameter "Intermediate" r "Salon" cylinders created technical oblems that provoked some genuine ngenuity. Various devices were fitted to cylinder phonographs to facilitate witching between the standard-sized nd "Inter" records, such as the well-ntrived British attachment installed on this Pathé "Le Sublime." The "auxiliary" reproducer carriage may e raised or lowered, as required, and is fixed in place by a thumbscrew. A convenient lever at the top of the device brings the reproducer stylus into contact with the record. (Value code: A)

2-65. Two 1 1/2" diameter tins, one holding glass reproducer diaphragms, the other a gooey preparation to stop phonograph belts from slipping. (Value code: G) *Courtesy of George Glastris.*

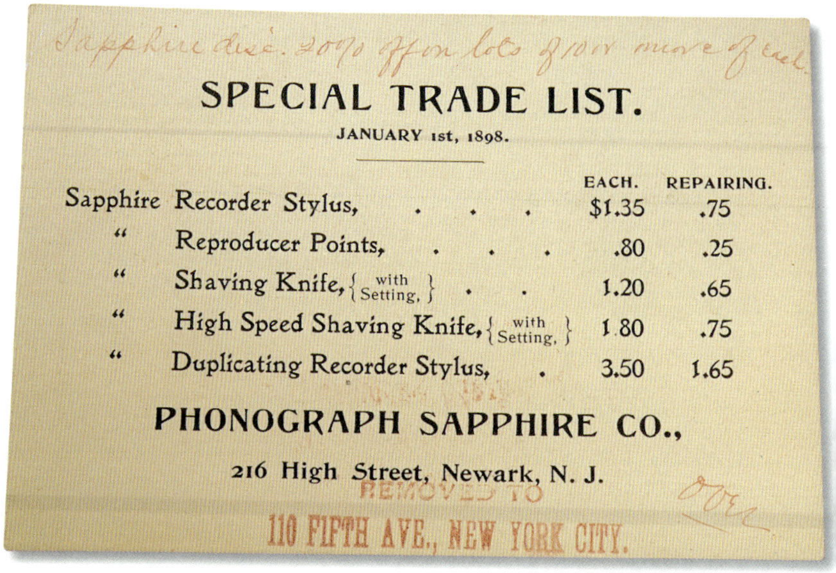

SPECIAL TRADE LIST.

JANUARY 1st, 1898.

	EACH.	REPAIRING.
Sapphire Recorder Stylus, ...	$1.35	.75
" Reproducer Points, ...	.80	.25
" Shaving Knife, {with Setting,} . .	1.20	.65
" High Speed Shaving Knife, {with Setting,}	1.80	.75
" Duplicating Recorder Stylus, .	3.50	1.65

PHONOGRAPH SAPPHIRE CO.,

216 High Street, Newark, N. J.

REMOVED TO

110 FIFTH AVE., NEW YORK CITY.

2-66. Although other materials were tried (including glass, "composition," metal and Carborundum), the sapphire became the standard of the industry for manufacturing cylinder phonograph styli. Recording and reproducing points, and sapphire "knives" for removing the grooves from a cylinder were sold by this specialized firm. (Value code: H)

2-67. Increasing the playing pressure of a talking machine reproducer resulted in greater volume (though increased record wear). Toward this end, a simple attachment was sold in Great Britain, consisting of a flat, nickel-plated metal plate to be fixed on top of the existing "fish-tail" counterweight of an Edison Model "C" reproducer. This attachment, stamped with three identifying music notations (sharp, flat, natural), may be seen in place here. *Courtesy of the Scott and Denise Corbett collection.* (Value code: H)

2-68. The Game of Uncle Josh was testimony to the widespread popularity of Yankee hayseed Uncle Josh Weathersby, impersonated on records by Cal Stewart. During the first decade of the twentieth century, Stewart's monologues made him the single most influential comedian in the recording industry. Although Stewart is frequently cited as having been the understudy to Denman Thompson, a well-known nineteenth century actor, there is evidence to suggest that this story was concocted to provide a pedigree. This uncomplicated parlor game was based on a premise familiar at the time: players recited from a text (in this instance resembling an Uncle Josh "laughing story"), inserting nonsensical phrases from the cards drawn at random. *Courtesy of Kevin Boerma.* (Value code: D)

2-69. Adjusting the speed of a cylinder phonograph, circa 1903, was a hit-or-miss proposition. Only a few years earlier, record speeds had not been standardized, and it was understood that the client would make minor alterations in pitch to his own liking. Beginning with the "New Model" Edison "Triumph" Phonograph (introduced 1901), the National Phonograph Company made an effort to mate the positioning of the speed control knob to a rudimentary scale indicating actual revolutions per minute of the mandrel. Other Edison Phonograph models, however, gave no particular indication of speed, and this must have troubled some more precisely minded owners. To the rescue came after-market gadgets such as the speed-indicating scale shown here on an Edison "Home" Phonograph Model "A." Once the speed had been calibrated (no doubt by using a handy paper disc stroboscope that could be affixed to the flat end of the mandrel) and the pointer set, the Phonograph operator could gauge the speed of the mandrel with relative accuracy. *Courtesy of Pat and Jamie Jones.* (Value code: D)

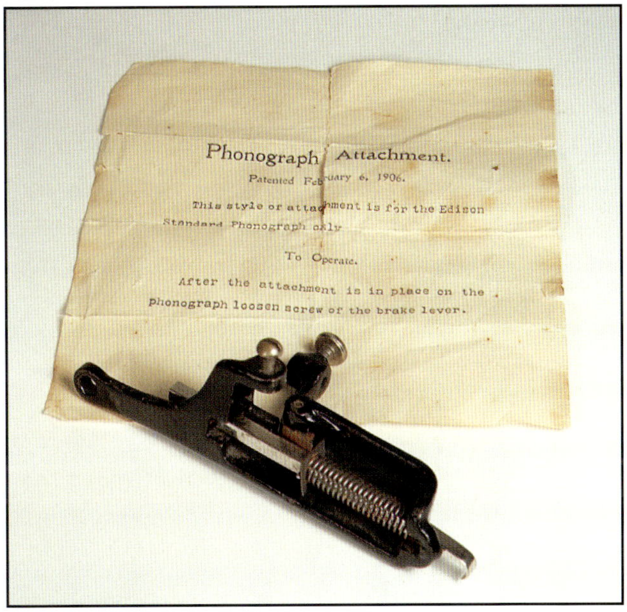

2-70. The necessity of returning to a phonograph to stop the motor after completion of a record was both an irritant and an inspiration to the human mind. Numerous mechanisms were devised to automatically shut off the instrument. This particular attachment, patented February 6, 1906, has been tipped on its side to show the spring-loaded mechanical finger which is the heart of its function. *Courtesy of the Charles Hummel collections.* (Value code: D)

2-71. The automatic stop is seen here secured in place using one of the "Standard's" mounting screws. The mechanical finger may be seen crooked around the brake lever, at the center of the illustration. As the Phonograph plays, the reproducer carriage tracks to the right. Eventually, the carriage contacts an adjustable metal rod (missing), triggering the finger to retract. The brake lever is thereby pulled to the right, stopping the motor. Zenas T. Grover was awarded United States patent No.811,633 for this device. *Courtesy of the Charles Hummel collections.*

2-72. A considerable variety of optional automatic stops were devised to equip cylinder talking machines, primarily Edison Phonographs. Graphophones were less popular with the inventors of auto-stops, but in some ways were more deserving of their attention. With the introduction of the Columbia "Twentieth Century" records (approximately 2" greater in physical length than conventional cylinders of the period) and the redesigned Graphophones to play them beginning in 1905, the owner of a late model Graphophone might play either the older or newer style records. An auto-stop for a Columbia in the "Twentieth Century" series (such as the "BF" or "Peerless" shown here) would require a wide range of adjustability. The mechanical shut-off installed on this machine has been set to the length of the conventional cylinders. As the carriage moves to the right, it nudges the slide at the front of the casting, elevating a finger that flips the brake lever to the left (as shown) into the "off" position. *Courtesy of the collection of Howard Hazelcorn.* (Value code: D)

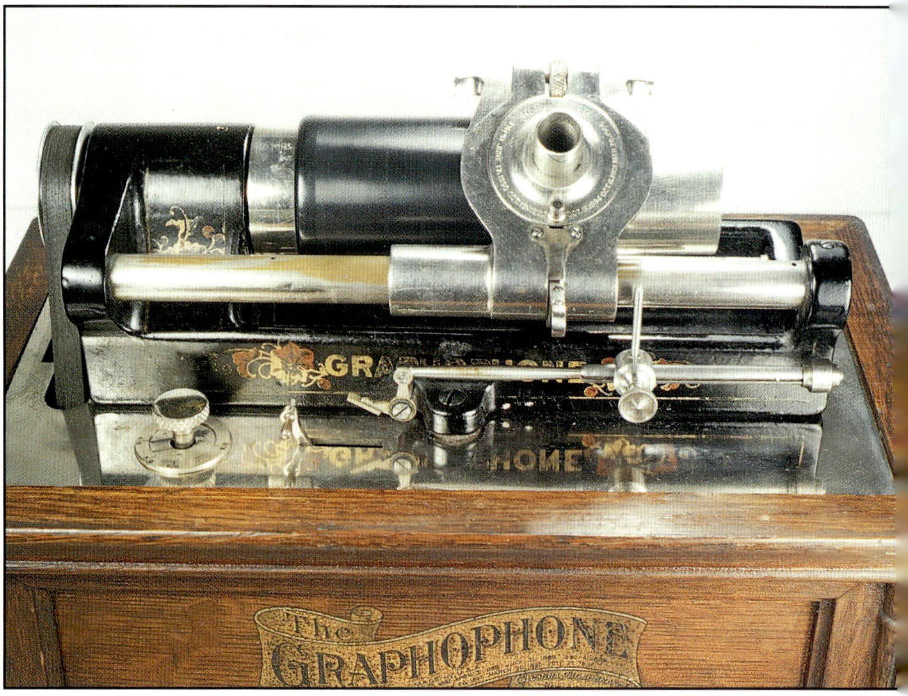

2-73. Edison dabbled with the convenience of having a reproducing and recording stylus attached to the same bar in the "Standard Speaker" of the early-to-mid 1890s. Alignment of the respective styli was achieved by modifying the position of the Speaker housing. Thereafter, Edison abandoned the principle in his domestic Phonographs, though, ultimately, this configuration would survive in office dictating equipment. Shown here, is a Columbia "Lyric" reproducer (introduced 1906) of a most peculiar stripe. Both a reproducing and a recording point are mounted on its needle bar, and the circular frame of the reproducer rotates within the diaphragm housing, allowing either stylus to contact the record. Clearly, it can not operate within the context of a conventional Graphophone such as the BK, BE or BF. However, this odd bit of technology challenges the imagination to conjure up the instrument with which it was originally mated. *Courtesy of the Babcock House Museum, John H. Perschbacher.* (Value code: VR)

2-74. The after-market automatic brake installed on this Edison "Standard" Phonograph Model "B" is more complicated than it first might appear. The complex alteration of the factory brake lever could benefit from some explanation. There are now *two* brake levers sandwiched one in front of the other. The one at the rear acts against the right surface of the governor friction plate to stop the motor. The one at the front works against the left surface of the friction plate, and is activated by the automatic stop arm. Here's how it works. Both brake levers are pushed to the left, starting the motor. The cylinder is played, and as the carriage tracks to the right, the adjustable threaded finger makes contact with only the *front* lever. It pushes the front lever to the right (the rear lever remains unmoved), which slowly brings a felt pad into contact with the left face of the rapidly spinning governor friction plate. This slows down the motor, and the governor relaxes, pressing even harder against the felt pad and stopping. The machine could also be operated conventionally by removing the threaded finger and using only the rear brake lever to control the motor in the manner of an ordinary Edison Phonograph. *Courtesy of the Scott and Denise Corbett collection.* (Value code: D)

2-75. The Edison "Standard" shown in the previous photo is seen here equipped with a very simply-designed volume control. The principle is that of a faucet: the knurled thumbscrew is tightened, limiting the flow of sound from the reproducer into the listening tubes. *Courtesy of the Scott and Denise Corbett collection.* (Value code: F)

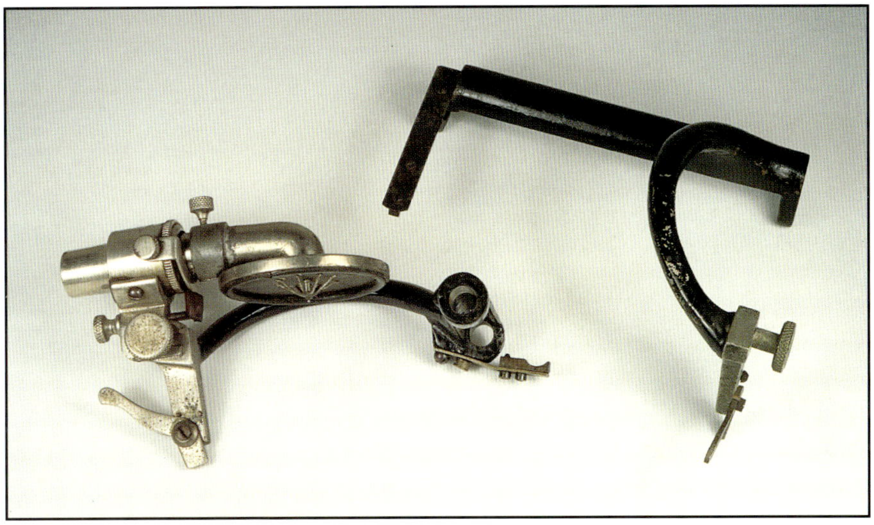

2-76. The Lambert Company of Chicago spawned a British cousin that was able to sustain itself after the American branch ceased activity. Of the two, only the British firm marketed a talking machine: the "Lambertphone." Lambert contracted its equipment manufacturing to the Excelsior Works of Germany, which also made "Lambert" attachments for refitting Edison Phonographs. Shown here is a carriage adapted to the Edison "Standard," supporting the Lambert "floating" reproducer. The "spider" connection of the stylus to the diaphragm was inspired by Bettini. To the right, is an adapter to be used with the Edison "Home" Phonograph. (Value code: C)

2-77. Around 1906, the British branch of Lambert marketed an attachment consisting of a carrier arm fitted with a Rawlinson sound-tight joint, and a Matthews floating reproducer featuring a diaphragm of larger dimensions than those of Edison. These attachments were available for "Gem," "Standard," and "Home" Phonographs. The Lambert device could be fitted to instruments of a significantly older vintage, such as this 1898-1899 Edison "Standard." *Courtesy of Gilbert (Bert) Pasley*. (Value code: A)

2-78. This flyer, circa 1906, advertised the Lambert conversion. Note the reference to the imminent introduction of attachments to fit Columbia Graphophones. That was a concept indeed like carrying coals to Newcastle. Whereas Edison's reproducer differed considerably from Lambert's, the one Columbia had employed for the previous ten years was virtually identical. Although redesigned Columbia models, featuring a reproducer more akin to Edison's, began to be released in 1904 (beginning with the "AZ") the idea of adapting Columbias to use floating reproducers is rather absurd. (Value code: G)

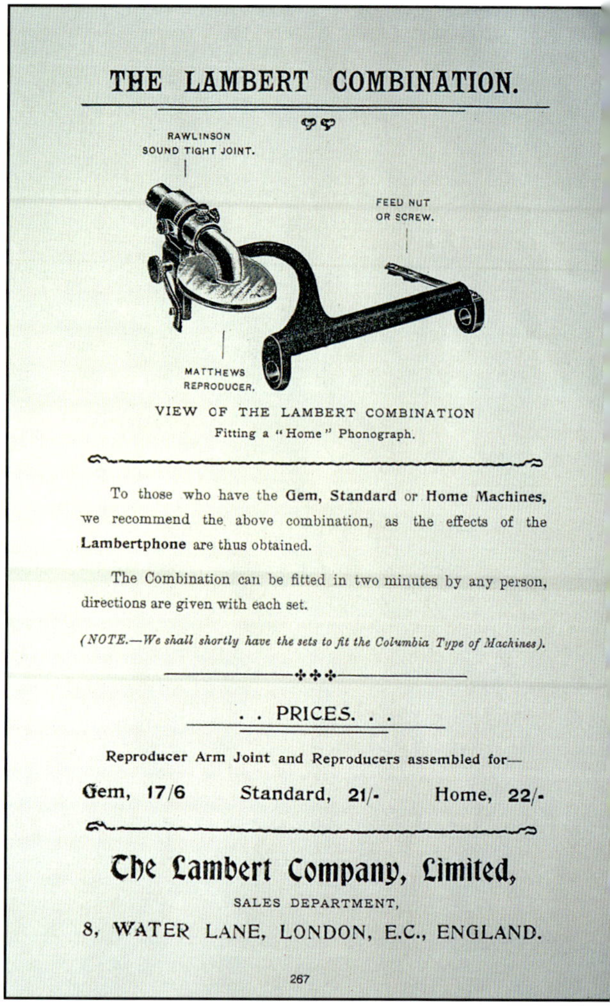

THE LAMBERT COMBINATION.

VIEW OF THE LAMBERT COMBINATION
Fitting a "Home" Phonograph.

To those who have the **Gem, Standard** or **Home Machines,** we recommend the above combination, as the effects of the **Lambertphone** are thus obtained.

The Combination can be fitted in two minutes by any person, directions are given with each set.

(NOTE.—We shall shortly have the sets to fit the Columbia Type of Machines).

. . PRICES. . .

Reproducer Arm Joint and Reproducers assembled for—

Gem, 17/6 Standard, 21/- Home, 22/-

The Lambert Company, Limited,

SALES DEPARTMENT,

8, WATER LANE, LONDON, E.C., ENGLAND.

267

2-79. Although coin-operated talking machines could be considered technologically sophisticated for their day, certain features were not commonly employed – such as a mechanical counter to keep track of the number of times the instrument was played. In the instance of this Edison "Eclipse" automatic Phonograph, circa 1906, an uncommon counting attachment may be seen mounted at the extreme right rear of the mechanism. The device is activated by a nickel-plated "finger" secured to the reproducer carriage. The mirror reflecting the Phonograph works was a common conceit in coin-ops, creating an illusion of depth, and allowing for a more impressive view of the mechanical wizardry. (Value code: VR)

2-80. Two flyers for interesting gizmos. The Morse "Clarifier" interposed a set of miniature "organ pipes" between the reproducing diaphragm and the amplifying horn of a cylinder or disc talking machine. The company never attempted to explain the exact technical function of this device, but testimonials apparently flooded in anyway. The 3 1/2" by 6 1/4" leaflet is brimming with the blind faith of Morse adherents. The "Tiz-It" was a metal ball-joint fitting that formed a connection between the reproducer and horn of a cylinder phonograph. The name was a take off of the phrase "This is it!" One good play on words deserves another, and Kreiling & Company which manufactured the "Tiz-It" also sold the "Shoz-It" speed dial and pointer to add a touch of precision to the external speed control knobs of cylinder talking machines. *Courtesy of Tom and Joanna Broome.* (Value code: H)

2-81. The Hooven Novelty Manufacturing Company created a simple device for testing the speed of a cylinder talking machine. After 1901, the revolutions per minute of the mandrel, which had previously varied according to company, recording engineer and a number of other factors, was standardized in the industry to 160. Since the "160" Speed Gauge was based on an interval of one minute, it would work equally well on instruments geared for either two or four-minute cylinders. Beyond this, the "160" Gauge was an advertising specialty, which could be stamped with the name of a merchant or distributor. *Courtesy of the Charles Hummel collections.* (Value code: F)

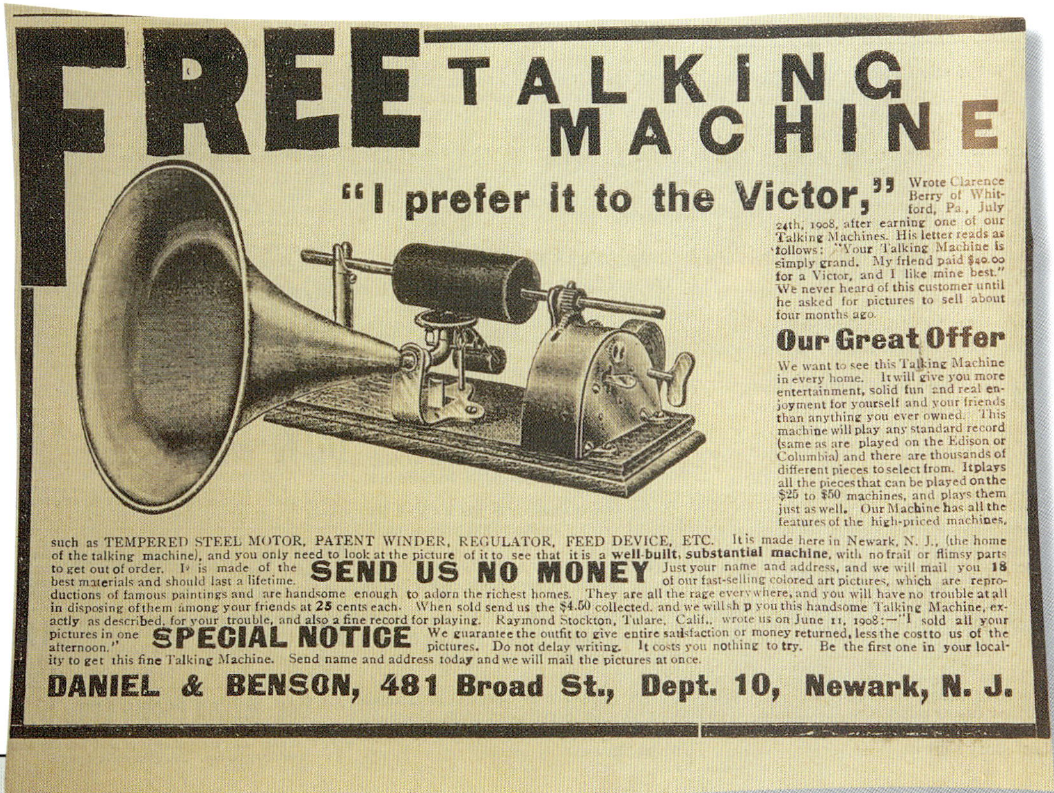

2-82. The "Rex" in basic design resembled another independently-produced instrument, the "Wizard." An advertisement in the September 1907 issue of *Talking Machine World* stated, "*Not an imported toy, but a low priced American Machine* that cannot get out of order… *Get it while it's new*… Made By Church Supply Company, 10 Barclay Street, New York." The "Rex" was a very modest cylinder talking machine in every way except for the advertising: a potshot at well-established Victor was pretty cheeky stuff. *Courtesy of Alan H. Mueller.* (Value code: H)

2-83. The Record Food Company of Kansas City, Missouri, sold a preparation in this jar for lubricating both cylinder and disc records. The customer applied "Record Food" before playing, and wiped it off afterward. What improvement of the sound quality or reduction of wear might have been obtained from this glop can only be hypothesized. *Courtesy of the Scott and Denise Corbett collection.* (Value code: F)

2-84. After the introduction of four-minute cylinders in 1908, Edison issued a "combination" attachment to allow existing "Gem" Phonographs to play the new records. No one could appreciate this more than the British, with whom the plucky "Gem" had been enormously popular. The mechanical bits shown here also adapted a two-minute "Gem" to dual functionality (British Patent No.2196, 1909). *Courtesy of George Glastris.* (Value code: D)

2-85. The ingenious British combination attachment on a "Gem" Model "B." One removes the middle gear and shaft of the upper drive train. The bearing adjusting screw from the left side of the feed screw is removed and used to secure the attachment in place. When the lever at the right is raised, the machine will run in the two-minute mode. When the lever is lowered, the instrument will play in the four-minute mode. *Courtesy of George Glastris.*

2-86. Edison issue[d] a set of operating instructions, printed on thin paper, pertaining t[o] the various models of cylinder Phonograph that the company sold. Subsequent developments in technology, such as the introduction of the four-minute records (which doubled the playin[g] time of the cylinde[r] format), required additional instructional material. (Value code: G)

2-87. Since Edison encouraged the owners of its Phonographs to convert their instruments to play the new four-minute records, it was inevitable that someone would market a tool to assist in snapping the pulleys of the "Spring-Motor" and similar models between the two and four-minute settings. The "pulley fork" seen here did the trick! *Courtesy of the Babcock House Museum, John H. Perschbacher.* (Value code: G)

2-88. Prior to 1906, the mainspring barrels of most Edison Phonographs were manufactured to be largely open on one face, in order to facilitate the replenishing of lubricating graphite. Shown here is a bottle of graphite supplied by Edison for this purpose. *Courtesy of Jean-Paul Agnard.* (Value code: G)

2-89. A 2 1/8" high container of "Finest Special Powder Graphite For Talking Machine Mainsprings... We recommend no other." *Courtesy of Mr. R. Chase.* (Value code: G)

2-90. The "Lewis Phono-Consonator" was sold (for $3.00) by the Lewis Manufacturing Company of New York City. When inserted between the reproducer and horn of a cylinder phonograph it supposedly "improves the tone without distorting the sound or decreasing the volume…eliminates all rattles, harsh and metallic grating, rasping nasal sounds and blasts and produces a soft, natural tone." A number of variations of this device have been seen, with slight differences of internal features based on a secondary vibrating (mica) diaphragm. Somewhat contradicting the ad copy, this particular "Consonator" has been equipped with a simple volume control valve. *Courtesy of Dan and Sandy Krygier.* (Value code: D)

2-91. For some reason, this Edison Model "N" reproducer was packed in an Edison cylinder record container instead of the conventional rectangular carton in which Edison reproducers and recorders were sold. One might be tempted to suggest that the blue label was transferred by a modern hand except that paper factory seal which protected the contents from tampering can be seen on both the lid and the body of the box. *Courtesy of Jen and Ken Flaherty.* (Value code: D)

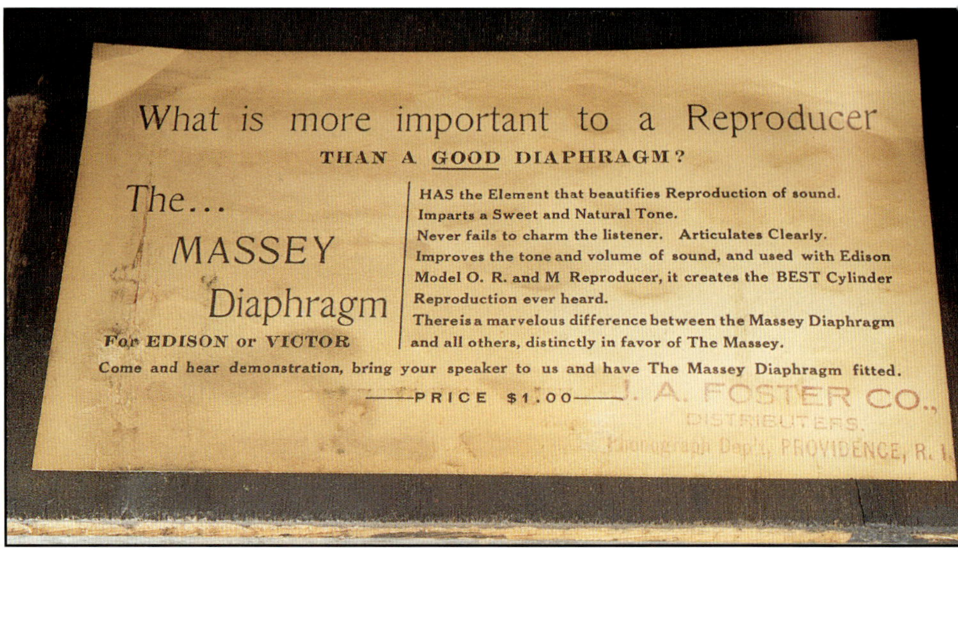
What is more important to a Reproducer

THAN A GOOD DIAPHRAGM?

The… MASSEY Diaphragm

For EDISON or VICTOR

HAS the Element that beautifies Reproduction of sound.
Imparts a Sweet and Natural Tone.
Never fails to charm the listener. Articulates Clearly.
Improves the tone and volume of sound, and used with Edison Model O. R. and M Reproducer, it creates the BEST Cylinder Reproduction ever heard.
There is a marvelous difference between the Massey Diaphragm and all others, distinctly in favor of The Massey.

Come and hear demonstration, bring your speaker to us and have The Massey Diaphragm fitted.

——PRICE $1.00—— J. A. FOSTER CO., DISTRIBUTERS. Phonograph Dep't, PROVIDENCE, R. I.

2-92. The lid interior of an Edison "Gem" Model "D." The "Massey Diaphragm" could be used with an Edison large-diameter reproducer or a Victor "Exhibition" soundbox (both diaphragms approximately the same size). Despite the purple praise heaped upon it by this flyer, we never learn what material the "Massey" was made of. This product and the "Massey Automatic Stop" were advertised by the J.A. Foster Company in the December 1911 *Talking Machine World*. "This little simple stop" for Edison Phonographs, that was "selling like 'wildfire'" worked in a rather direct fashion: when the reproducer carriage reached an arbitrary spot near the end of the record, it "stalled out" against the frame of the motor casting, bringing the machine to a grinding halt! *Courtesy of Skinner's Auctions.*

2-93. A stately mahogany cabinet conceals a most unusual Edison Phonograph, modified to perform far beyond its original specifications. Discovered in Canada the early 1970s, this machine has resided in three private collections during the intervening years, and continues to elude definitive explanation. *Courtesy of the Scott Malawski collection.* (Value code: VR)

2-94. Removing the lid reveals an Edison "Fireside" Model "A" (introduced 1909). The reproducer is an odd, unmarked apparatus employing a diaphragm of uncommonly large dimensions (and not originally found with this particular machine). Barely visible at the rear of the machine is a repeating mechanism, which lifts the carrier arm (aided by a counter-weight mounted to the back of the carrier arm) and returns it to the beginning of the selection. Sharp-eyed collectors will have also noted a speed control knob mounted *above* the cast motor bed-plate (unlike the customary "Fireside" control). However, of all the peculiarities, perhaps the most noteworthy is to be found *within* the cabinet. *Courtesy of the Scott Malawski collection.*

2-95. The original single mainspring of the "Fireside" (left) has been supplemented by the three additional mainsprings shown at right, creating a leviathan motor of extraordinary power. The most likely hypotheses for the machine's intended use are either to have continuously repeated an advertising record or to perform a prolonged presentation of music for dancing. The tasteful cabinet and professional custom mechanics seem to support the latter theory, suggesting a domestic use – perhaps during the American dance craze of the early teens. *Courtesy of the Scott Malawski collection.*

2-96. Introduced in 1910 as an accessory for the Edison Amberola "1-A" Phonograph, the Model "M" reproducer is usually seen in two variations. The earlier version employs a "circular" counterweight similar to the one shown here. Later on, the weight was reduced in size to a more triangular shape, in order to control record wear by decreasing the pressure of the styli on the cylinder. What has not been seen before, however, is this anomalous style of weight, terminating in a "fishtail" that much resembles the design of the well-known Model "C" or "H" reproducers. The added mass of *this* weight would have produced excellent volume, but would have compromised the integrity of the records. *Courtesy of the Aaron and Thea Cramer collection.* (Value code: VR)

2-97. The method of engaging the reproducer carriage of an Edison Phonograph illustrated here, employing a finger-friendly slide instead of a lever or a knurled plunger, is sometimes seen on the "Triumph" Model "D-2," introduced in late 1909. This particular instrument has been custom nickel-plated. Despite the fact that the carriage slide is clearly an improvement, most versions of the "Triumph" kept to the old-fashioned lever which had been in use since the 1890s. *Courtesy of the Sanfilippo collection.*

2-98. Among its wares, talking machine distributor Louis Buehn & Brother of Philadelphia promoted these two brands of machine oil. "Many Use" was a coined name, such as the more familiar "3-in-1." Nothing, however, could be more expressive of ameliorative effects than "Phono-Vita." Flyers, 3 1/2" x 6 1/4". *Courtesy of Tom and Joanna Broome.* (Value code: G)

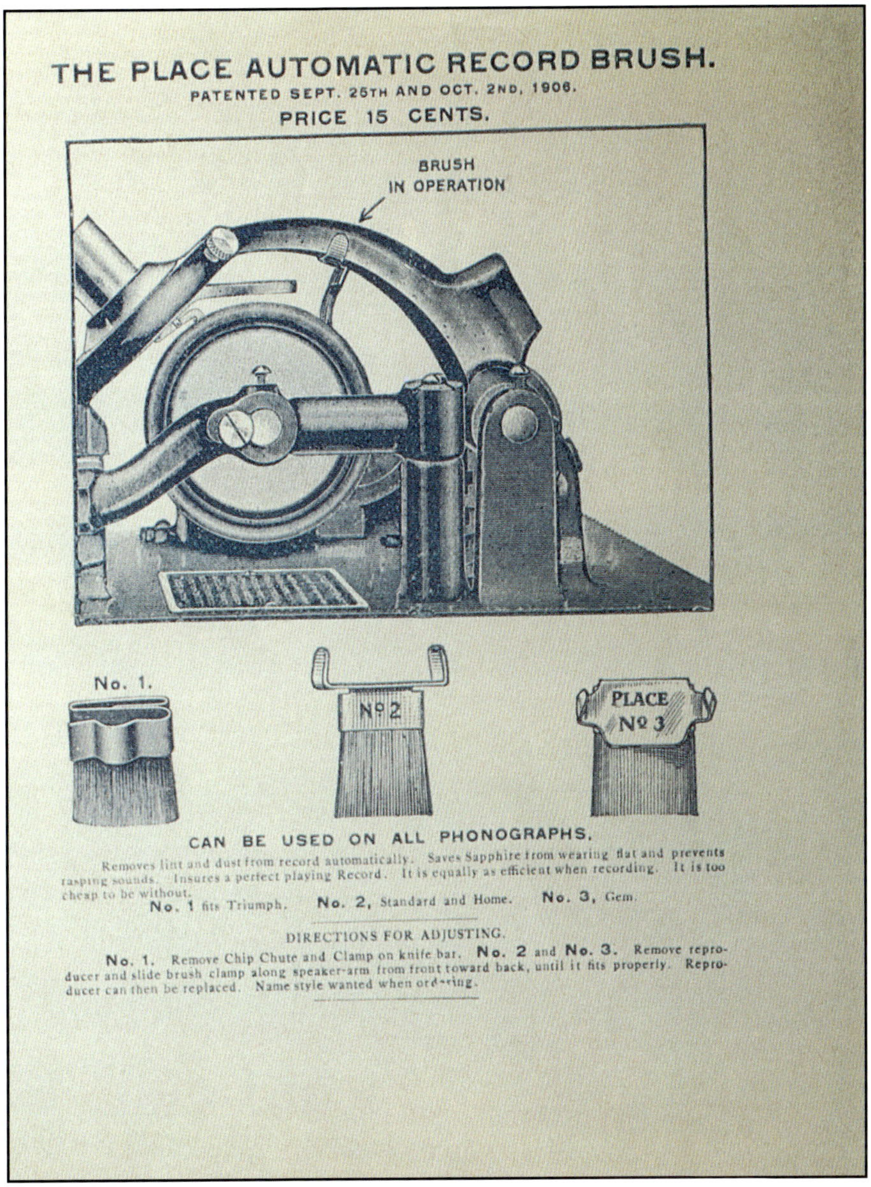

THE PLACE AUTOMATIC RECORD BRUSH.

PATENTED SEPT. 25TH AND OCT. 2ND, 1906.

PRICE 15 CENTS.

BRUSH IN OPERATION

No. 1.

Nº 2

PLACE Nº 3

CAN BE USED ON ALL PHONOGRAPHS.

Removes lint and dust from record automatically. Saves Sapphire from wearing flat and prevents rasping sounds. Insures a perfect playing Record. It is equally as efficient when recording. It is too cheap to be without.

No. 1 fits Triumph. **No. 2,** Standard and Home. **No. 3,** Gem.

DIRECTIONS FOR ADJUSTING.

No. 1. Remove Chip Chute and Clamp on knife bar. **No. 2** and **No. 3.** Remove reproducer and slide brush clamp along speaker-arm from front toward back, until it fits properly. Reproducer can then be replaced. Name style wanted when ord-ring.

2-99. Place "Automatic Record Brushes" came in various configurations, to clean cylinders (or discs) as they played. These uncomplicated gadgets clipped to a part of the talking machine mechanism, such as illustrated on this flyer, 5 1/4" x 6 3/4". *Courtesy of Tom and Joanna Broome.* (Value code: G)

2-100. Shown here are two styles of Place record brush, for cylinder phonographs. Henry A. Place, Frans E. Forsell and Joseph W. Blackman received the patents for the record cleaning brushes that would be manufactured and sold by the Blackman Talking Machine Company of New York City. As a 1911 advertisement stated, "[The Place brush] AUTOMATICALLY CLEANS record grooves, insuring a smooth track for sapphire or needle. Reduces friction to minimum. Enables needle to wear better and play good all through the record. Keeps sapphires from wearing flat." By the mid-teens Blackman had improved the brand name of these brushes to "Cleanrite." (Value code: H)

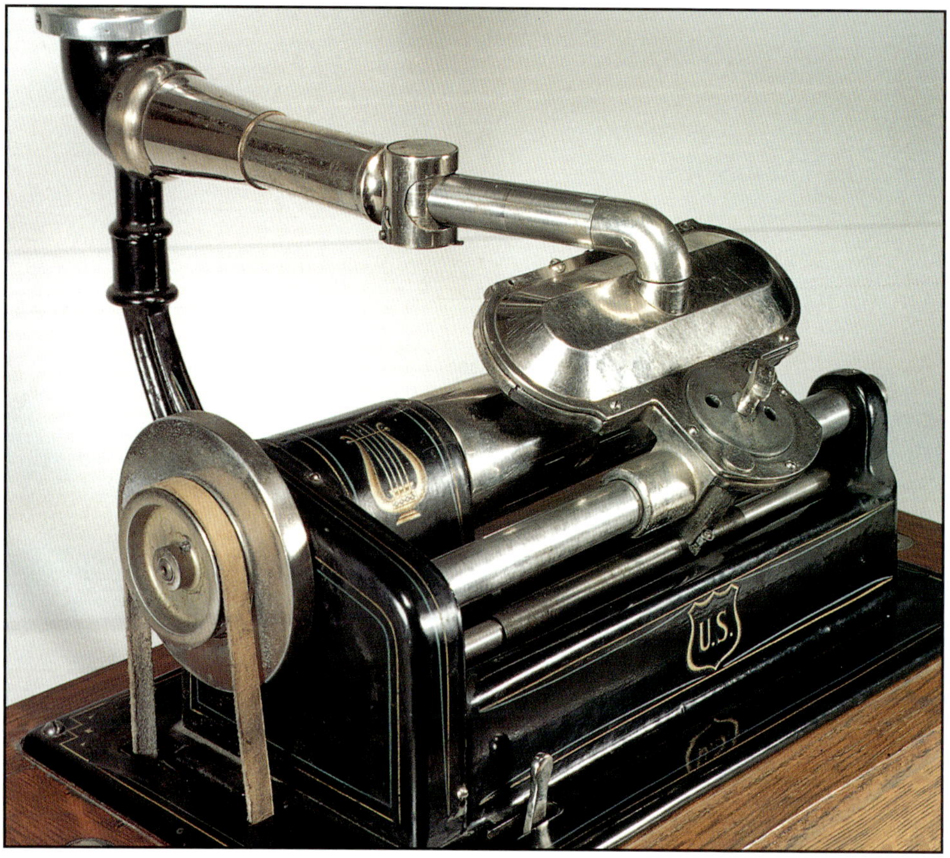

2-101. The U-S Phonograph Company of Cleveland, Ohio sold a fascinating line of cylinder talking machines 1910-1913. This firm combined the patents of Harry McNul and Thomas Towell for a complex two-and-four-minute talking machine with the pate of Varian Harris for a cylinder record consi ing of a thin celluloid skin over a compositi core. Production resources were provided b the Babcock-Bishop-Becker Company (a so fountain manufacturer). The most common seen fashion in which the U-S Company connected its formidable die-cast reproduce to the "back-mounted" horn was a tapered, segmented, articulated arm. The earliest production, however, was quite different. A seen here, the fixed arm assembly much resembled the type of connection that Ediso would adopt for its Amberola "V" (1912). In the September 1910 *Talking Machine World*, U-S Company announced its "New Flexible Tone-Arm," replacing the configuration pictured here. "By means of this device the reproducer carriage is given a free range of travel… Not only this, but the Flexible Ton Arm secures a **Greater Amplifying Radius** the sound waves, and brings out the detail i way that is truly remarkable." *Courtesy of Je Paul Agnard.* (Value code: VR)

2-102. The Edison "Kinetophone" system of sound-synchronized motion pictures was exhibited 1913-1916. The Kinetophone involved a Phonograph and a movie projector (both electrically driven), an amplifying horn, film and cylinder records. The records were of unique size, measuring 4 3/8" diameter by 7 3/8" long. Here, a "Kinetophone" cylinder for the film "College Days" is shown next to a conventional Blue Amberol record of the same period. *Courtesy of the Charles Hummel collections.* (Value code: VR)

2-103. No, you haven't seen this in the foyer of the Museum of Modern Art – it was once a cylinder record mold. This artifact was dug out of the ashes of the fire that devastated a considerable portion of Edison's factory complex, in 1914. *Courtesy of Gilbert (Bert) Pasley.* (Value code: VR)

2-104. In 1915, Edison introduced a simplified line of "Amberola" Phonographs, consisting of one upright and two table models. This pamphlet, from the year of their introduction, also illustrates the "V" (introduced in 1912), which the company was trying to close-out.

The design of the 1915 "Amberolas" featured an internal horn suspended within the cabinet. Experience has proved that this horn is likely to collect dust. The paper tag shown here, a truly ephemeral object, advised the client about keeping his Amberola neat. (Value code: G)

2-105. Giving up on something is not a British trait. So it was with Edison's cylinder record in the late 1920s – many, many Americans didn't know that cylinders were yet being produced, but the Brits kept their old Phonographs playing. Even in the mid-1930s, after Edison cylinder production had ceased, British physicist Adrian Sykes found a small market for his "Electrograph." This device, employing the floating weight of an Edison "Diamond B" reproducer coupled with an electric pick-up, allowed the owners of Phonographs with a "horizontal" carriage to play Edison Blue Amberols through a radio set. *Courtesy of George Glastris.* (Value code: C)

Three:
Disc Talking Machine Gadgets

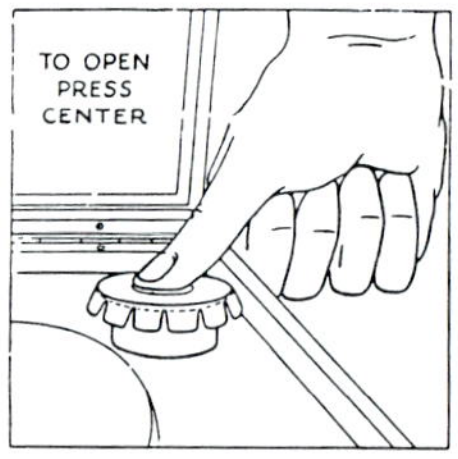

Unlike cylinder talking machine accessories, the marketing history of which was relatively short, devices for "improving" the performance or appearance of disc machines emerged, proliferated, and endured for decades. Although the authors have usually arranged artifacts in approximate chronological order throughout this and other books, this chapter must be an exception. The longevity of certain products, along with the mysterious pedigrees of others, would make a sequential presentation confusing and of limited value. Rather, items have been generally grouped relative to function, origin, or other relationships, which are sometimes not obvious. In each chapter, you will find a few items that are more fully represented in one of the other chapters. Our purpose is to emphasize how artifacts within general fields of definition "cross over." For instance, carrying cases can be considered talking machine accessories as well as units of storage – often for both machines and records. We hope, however, our methods will not provoke madness.

Those using this chapter as a research tool are encouraged to employ the index to locate particular items. Others, casually perusing these pages for pleasure, will share one part of the experience of the early twentieth-century buying public: facing a dizzying kaleidoscope of accessories and contraptions. Collectors, passionately examining these pages, will encounter the other part: surveying a host of tantalizing objects ultimately designed to open the purse or wallet of the prospective customer.

3-1. An idealized representation of "modern" technology, featuring a 23 1/2" high androgyne cradling a disc talking machine. The left hand would have held some symbolic object such as a laurel wreath. The brass tag reads, "La Rénommée Couronnant le Progrès, par Delavigne." (Progress crowned by success, by Delavigne). *Courtesy of Jalal and Charlotte Aro.* (Value code: VR)

3-2. Art nouveau flourishes flank this graceful decal adorning a European disc talking machine. Among the "modern inventions" purveyed were Gramophones from England and France, Zonophones and Odéons from Germany, as well as French-made Dutreih cylinder phonographs. *Courtesy of Jalal and Charlotte Aro.*

3-3. A variety of gimmicky record dusters were available over the years. Shown are three examples. At the top is a "Dustix" Record Cleaner, made by the E.T. Gilbert Manufacturing Company. The user would hold this device in place for a few seconds, while the record revolved on the turntable. The holes were to accommodate 10" or 12" records. At lower left is a 3 1/2" diameter duster with a "used needles" cup built into the center. To the lower right is a 4" long "Dulcetto" magnetic cleaner, which would supposedly remove microscopic shards of steel needles from the record grooves, or assist the owner in retrieving stray needles from around the talking machine mechanism. *Courtesy of Walter and Carol Myers.* (Value code: G)

3-4. The Boye Needle Company of Chicago merchandised both sewing needles and phonograph needles. The company also offered "High Grade 'Boye' Oil" as attested to by the 6" high bottle in the center. On either side are containers for "Eureka Noiseless Talking Machine Lubricant," which was a grease rather than oil. (Value code: G)

3-5. One of the most intriguing disc gadgets ever conceived (although sadly not manufactured) was this design by James Vincent Cruso, for which a United States patent was granted on August 28, 1906 (No. 829,848). As the disc talking machine gained popularity during the first decade of the twentieth century, several inventors sought to design an apparatus which would allow both cylinder and disc records to be played on the same machine. The Type "HG" ("Home Grand") Graphophone shown here is one of two known survivors (both "Home Grands"), which were professionally modified using Cruso's pattern, possibly by the American Graphophone Company. In any event, nothing came of the experiment other than this fascinating hybrid relic. *Courtesy of Douglas DeFeis.* (Value code: VR)

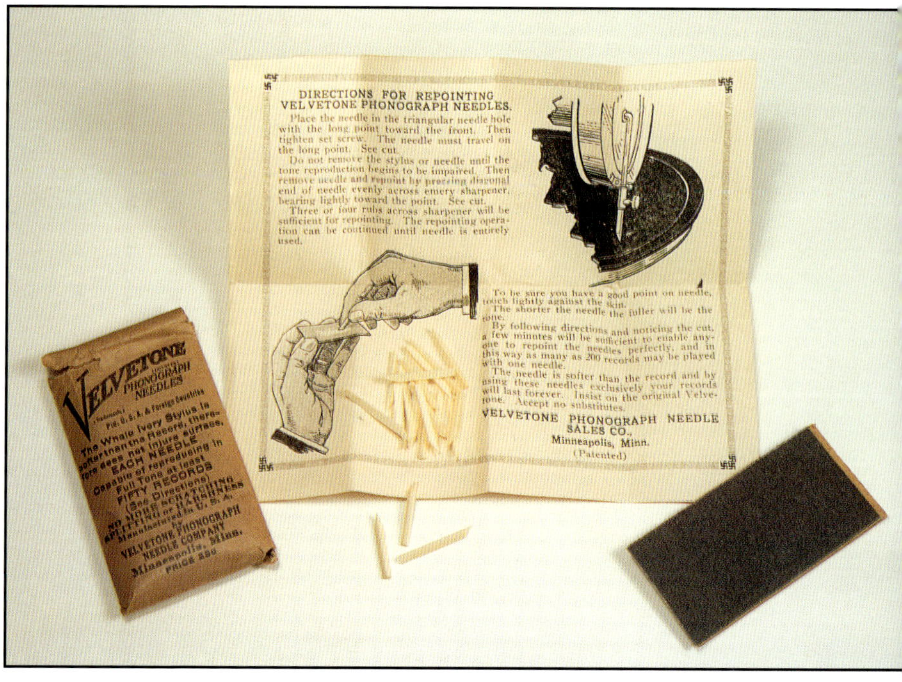

3-6. There were, of course, steel phonograph needles, fiber needles, bamboo and cactus-spine needles. More unusual are "whale ivory" needles, shown here with an emery sharpener and directions for use. Note the ancient good-luck symbol, the swastika, in the corners of the direction sheet. Ivory needles were advertised in the *Talking Machine World* in 1908 – an ad for Baldwin Ivory Needles stated, "...each Needle may be used several times without change, and when worn the point may be easily renewed on the Renewing Block, furnished free." *Courtesy of Donald Walls.* (Value code: G)

3-7. The Tusko Manufacturing Company of Chicago produced a full range of talking machine accessories. Shown here is a packet of Tusko needles. The firm was long on enthusiasm; a bit short on copy-writing talent: "Tusko talking machine needles are a wonderful innovation as they are practically indestructible and are made of a natural substance more nearly the same temper of a record than anything heretofore discovered… One Tusko needle will play hundreds and hundreds of times and render better music, they are the cheapest needles you can buy." Tusko needles were advertised in the *Talking Machine World* from May-August 1909. *Courtesy of Tom and Virginia Hawthorn.* (Value code: G)

3-8. The "Everplay" was a so-called "permanent" needle which allowed the user to replenish the point by feeding out a tiny steel wire whenever necessary. The $1.00 price must have seemed high to buyers accustomed to steel needles of the cheap, throwaway variety. *Courtesy of Donald Walls.* (Value code: F)

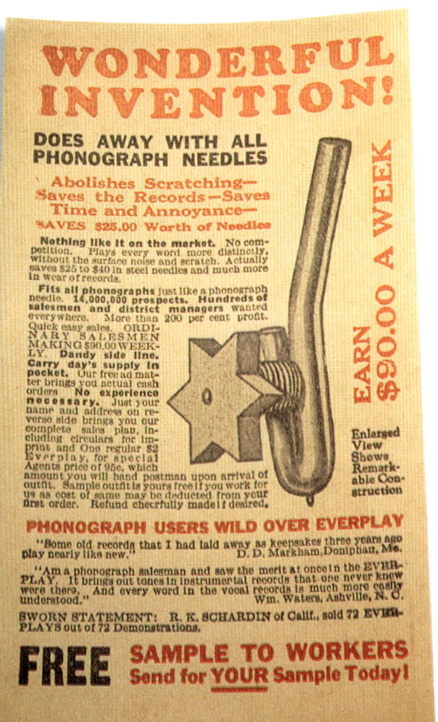

3-9. A postcard for the "Everplay" pushed it as "a dandy sideline." *Courtesy of the Scott and Denise Corbett collection.* (Value code: H)

3-10. An original dealer's box of Pathé sapphire styli. The Pathé disc record system, promoted heavily in the United States during the teens, was a "vertical cut" format. Sound vibrations recorded at the bottom of the record groove were picked up by a permanent sapphire ball set in a brass shank, such as is contained in each little envelope. (Value code: G each)

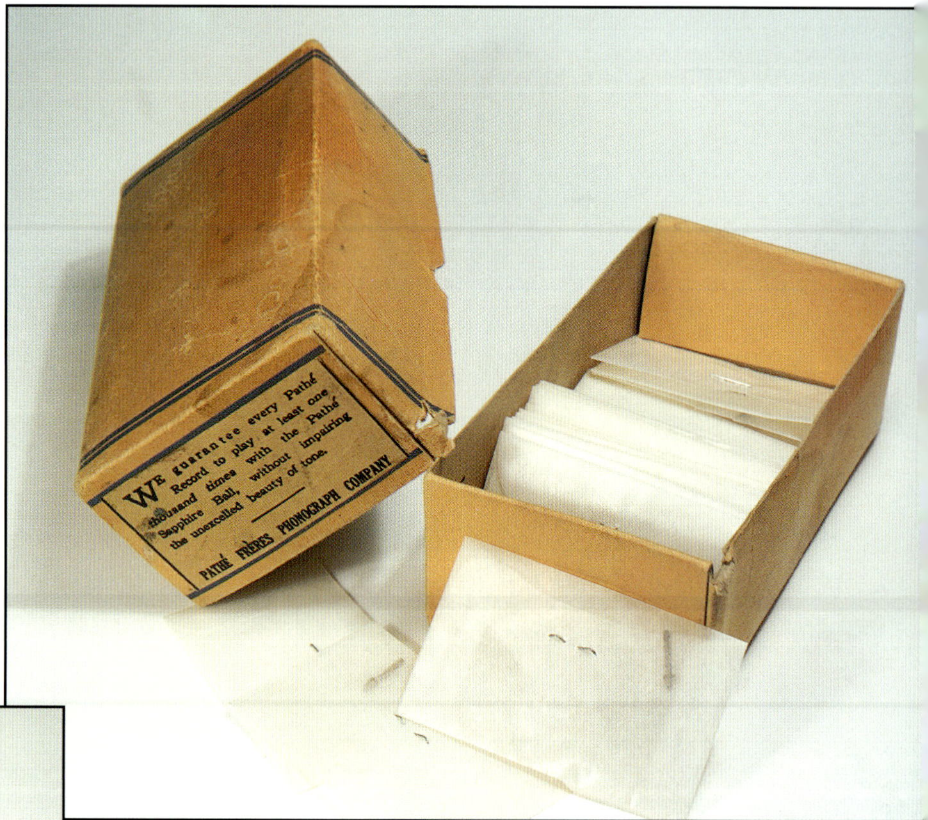

3-11. "Golden Pyramid" was a uniquely-packaged brand of needle popular in Great Britain. Inverting the gleaming tin dispensed one needle at a time. It has not been proved that storing needles in the mystical atmosphere of a pyramid improves the sound of a record, but it is at least as effective as placing a magnet under your talking machine. *Courtesy of the Scott and Denise Corbett collection.* (Value code: VR)

3-12. The steady demand for talking machine needles represented a lucrative adjunct to the sales of machines and records. Needle manufacturers kept phonograph dealers apprised of the newest products and display materials. Shown are 3 1/2" x 6 1/2" flyers for "Okeh," "Nupoint," and "Reflexo" needles, all distributed by the Consolidated Talking Machine Company. The latter two claimed the ability to play ten records per needle, probably to the detriment of the discs. The craft of thinking up snappy names for needle brands was what dreaming up clever names for hair salons is today. *Courtesy of Tom and Joanna Broome.* (Value code: H)

3-13. The American Phonograph Needle Company (note the printed "NOT INC.") supplied this sample package for dealers to publicize its "Musical Needle" and its accompanying sharpener. To those who wondered about the substance from which the needles were made, the reverse side of the envelope confided: "This new needle is made from the heart of a tropical plant from South America. It is tempered by a secret process known only to the Inca Indians and used by them to harden their arrow points – that is why it plays so well and is so durable." The "tropical plant from South America" was probably nothing more than cactus spines from the southwestern United States. *Courtesy of Charles McCarn*. (Value code: H)

3-14. This little 1 3/4" high tin contains Royal Blue Certified Needles, manufactured by the F.C. Henderson Company, Boston, Massachusetts. The motif was inspired by popular chinaware. *Courtesy of Charles McCarn*. (Value code: G)

3-15. An unusual geared cutter for fiber needles. Marked "Lidseen Needle Cutter." *Courtesy of Charles McCarn.* (Value code: F)

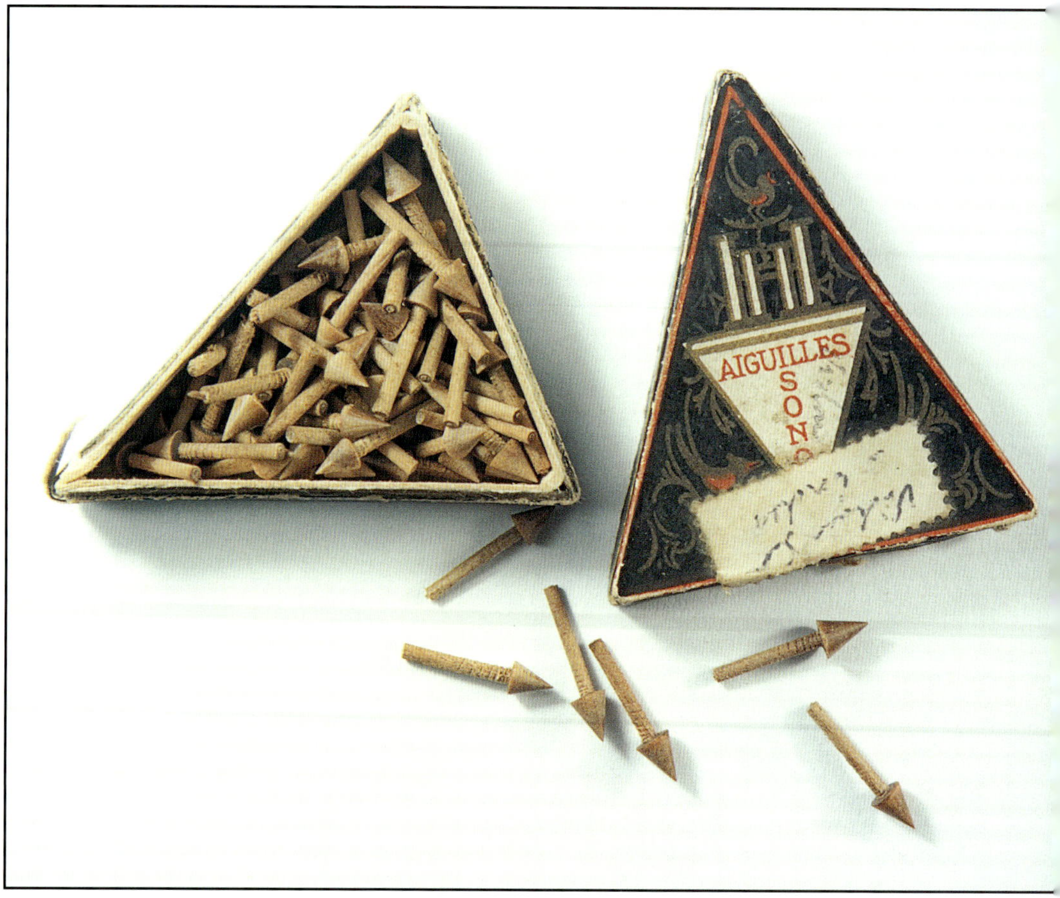

3-16. Peculiar box; peculiar needles. A triangular container housing oddly shaped "Sonor" fiber needles. *Courtesy of Yves Rouchaleau*. (Value code: F)

3-17. This Odeon "Electric" needle tin, 1 3/8" x 1 7/8" has a convenient celluloid window on its underside, through which the paper, inner protective sleeve can be seen, or the status of its contents checked. *Courtesy of the Charles Hummel collections.* (Value code: F)

3-18. An automated needle tin? Why not – this German "Pegasus" tin hides a clever secret. *Courtesy of the Charles Hummel collections.* (Value code: D)

3-19. Squeezing the two halves of the "Pegasus" tin together activates an arm, which advances a spiral carousel loaded with steel needles. One needle at a time is dispensed – blunt end first for safety! *Courtesy of the Charles Hummel collections.*

3-20. A tin within a tin within a tin. Beka records and needles were German-made for the British market. The 200-count tins fit into a larger tin, a number of which can be stored in the largest container shown. *Courtesy of the Charles Hummel collections.* (Value code: G each)

3-21. A box of chocolates, perhaps? These self-possessed dancers, circa 1850, give no hint of what might be contained in this tin, 2 5/8" high, 7 3/8" deep and 11 1/8" wide. It in fact holds containers of Burchard (German) steel needles. *Courtesy of the Charles Hummel collections.* (Value code: VR)

3-22. "Peacock" (Japanese) brand tins of needles were held in this colorful container, 2 1/4" high, 5 1/4" deep and 6 1/4" wide. *Courtesy of the Charles Hummel collections.* (Value code: VR)

3-23. Children's records became increasingly popular during the 1920s. Often seen are examples of the popular Columbia-recorded "Bubble Book" series, but less common are corresponding Victor products (which were produced at different times). Shown is a 7 3/4" x 9 1/2" book of recorded nursery rhymes bearing the familiar Victor label. *Courtesy of Charles McCarn*. (Value code: F)

3-24. Another well-known children's record label from the 1920s was "Little Tots." Shown are three albums of "Little Tots" 7" discs and stories. *Courtesy of Walter and Carol Myers.* (Value code: F)

3-25. An assortment of children's records including the Harper-Columbia ("Bubble Book") label, "Songster," "Cameo-Kid," and "BobOlink." *Courtesy of Martin F. Bryan*. (Value code: H)

3-26. Another gimmick to boost music store business was the Talking Book, a recording attached to a colorful die-cut picture ("the kids bring their mothers, and you can sell other records also"). These colorful cardboard images, such as animals, were marketed by the Talking Book Corporation of New York City. The size was kept within the limits of a typical 12" turntable. Talking Books were manufactured under license from the Emerson Phonograph Company, as announced in the May 1919 issue of the *Talking Machine World.* These items received hard use from children, and were often damaged due to their delicate shapes. *Courtesy of Steve and Ellie Saccente.* (Value code: E)

3-27. Another Talking Book Corporation product of circa 1919 in the impersonation of a frog. New designs were introduced each month. "Stock Talking Books for the biggest Christmas demand in years [1919]! The public is going to make up for the heartache of war time by blowing the lid higher than a kite!" *Courtesy of Norm and Janyne Smith.* (Value code: E)

3-28. The reverse of the above gives an idea of the record's content. *Courtesy of Norm and Janyne Smith.*

3-29. Four-legged animals such as this fox were part of the series... *Courtesy of Charles McCarn.* (Value code: E)

3-30. ...As was this circus elephant. Most Talking Books sold for 25 cents. "Discount [to dealers] is 1/3 off and over, according to quantity. Don't lose another day of Talking Book profits! Order today!" *Courtesy of Charles McCarn*. (Value code: E)

3-31. These Tuck postcards, circa 1910, were an English innovation designed to add some novelty to the genre. Of incidental concern to the company was the possibility that the receiver of the postcard might have to purchase a talking machine. (Value code: H each)

3-32. A 5" x 6" "Greetaphone" card. These cards were sold with recordings appropriate for various occasions. Marketed by Hall Brothers, Kansas City. *Courtesy of Steve and Ellie Saccente*. (Value code: G)

3-33. The permutations of Edison-sponsored peripherals were almost endless. Shown are scarce and common examples. At left is a bottle of Edison Phonograph Cabinet Polish, an unusual item. Far more common are record dusters similar to the one at right. More frequently encountered items such as dusters offer a virtually unlimited variety to the collector – they came in many styles, and often carry the name of a former dealer in the collector's hometown. *Courtesy of Norm and Janyne Smith*. (Value code: F)

3-34. In the post-World War I talking machine market, many manufacturers included conversion apparatus adaptable to both vertically and laterally recorded discs. This might include a swiveling, jointed tone arm, or supplemental parts, as employed by the "Silvertone." Among these thoughtful merchandisers was the Cheney Company, which included this elbow, sapphire ball (for Pathé) and diamond stylus (for Edison). (Value code: F)

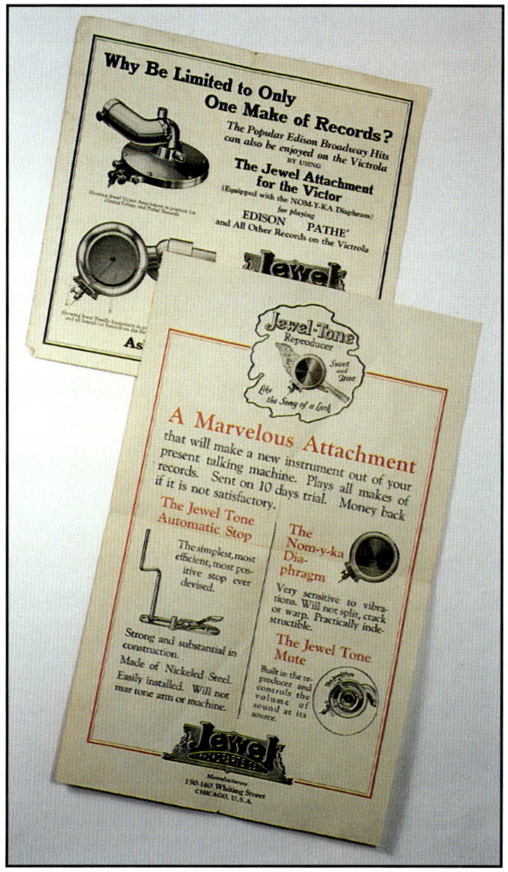

3-35. The Jewel Phonoparts Company of Chicago must have roiled with activity around 1916. It was marketing a variety of talking machine accessories, especially those suited to converting between the various lateral and vertical cut disc record systems then available. In these brochures we see offered everything from an automatic stop to a volume control. (Value code: H)

3-36. The "Claravox" was one of several devices available to allow the playing of Edison's specialized, vertical cut Diamond Discs on a competitor's phonograph. The use of a cork stiffener on the diaphragm imitates the design of Edison's own equipment. (Value code: D)

3-37. The Union Specialty and Plating Company of Cleveland, Ohio, was well-known for its lateral/vertical record conversion attachments for Diamond Disc Phonographs. Shown is a device that allowed Pathé style vertically cut discs to be played on a Columbia Grafonola by using a sapphire ball. (Value code: D)

3-38. A more unusual approach to playing vertically recorded discs was this unmarked attachment (resembling the "Edephone" below), which hearkened back to Columbia reproducer designs of the 1890s. This device was also meant for use on a later Disc Graphophone or Grafonola. (value code: D)

3-39. One of the disadvantages of purchasing an Edison Diamond Disc Phonograph was the steadfast refusal of Edison to endorse the playing of Victor and Columbia (lateral) records. A variety of attachments were marketed to meet that need. Far less common was the owner of a Victrola or Grafonola who yearned to hear Edison discs on his machine. Nevertheless, an enterprising manufacturer in Los Angeles, California offered an accessory (the "Edephone") which would enable the vertically recorded Edison records to be played on virtually any disc phonograph. *Courtesy of the Scott and Denise Corbett collection.* (Value code: D)

3-40. The "Edephone" attachment, playing an Edison Diamond Disc on a Victor "V." *Courtesy of the Scott and Denise Corbett collection.*

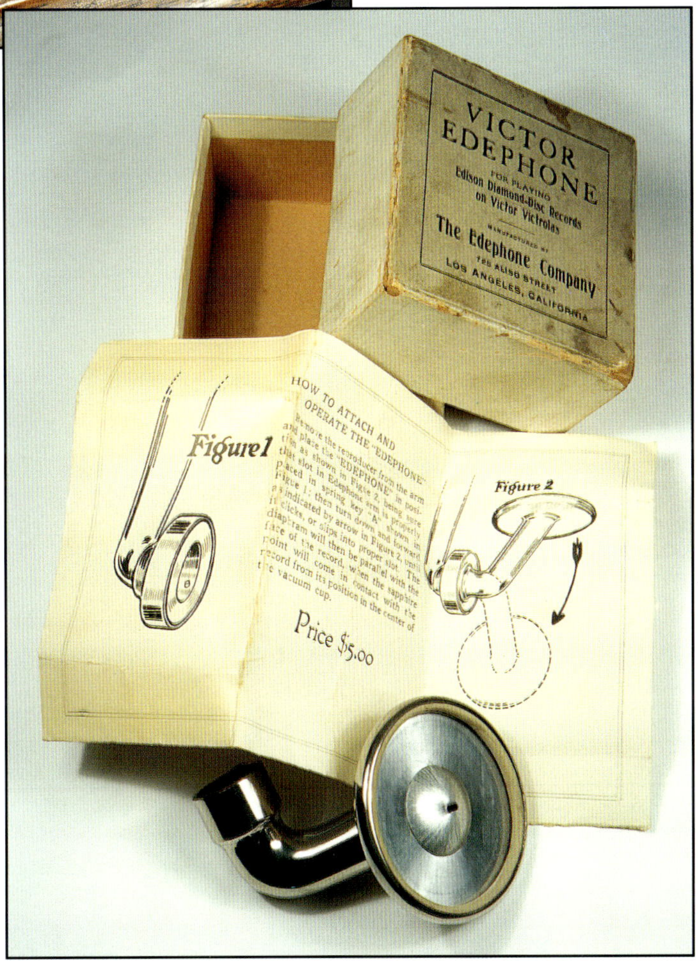

3-41. The discovery of an "Edephone" attachment with its original packaging and instructions contributes further points of information. Note that the box reads "Victor Edephone for playing Edison Diamond-Disc Records on Victor Victrolas." Yet, the instructions clearly illustrate and describe the tone arm of a Columbia Grafonola. One wonders if the company saved money by packaging all models in the same style box. *Courtesy of the Babcock House Museum, John H. Perschbacher*. (Value code: D)

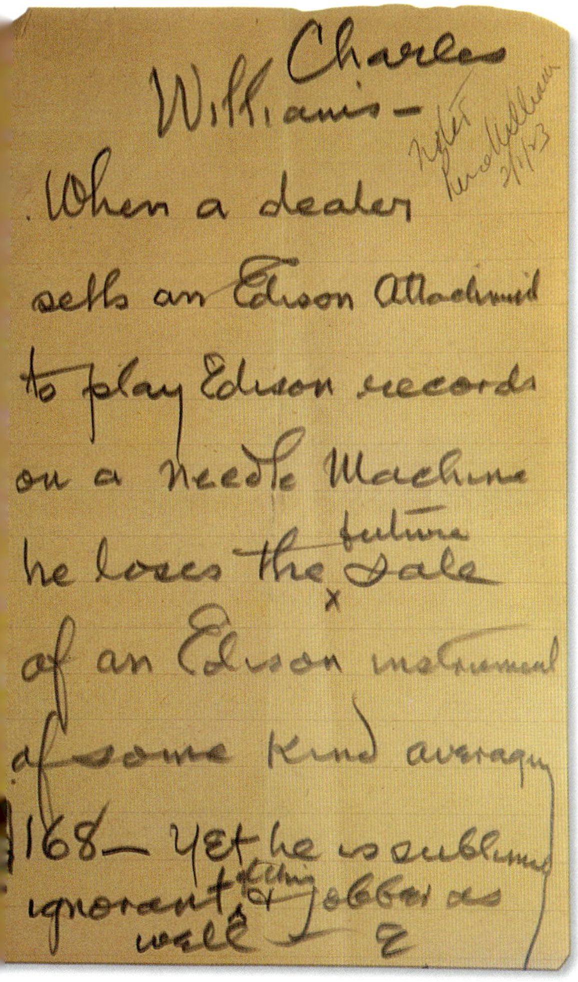
Charles Williams –
When a dealer sells an Edison Attachment to play Edison records on a needle Machine he loses the future sale of an Edison instrument of some kind averaging 168 – yet he is sublimely ignorant of this & jobber as well
E

3-42. Any question of how Thomas Edison himself felt about using attachments to play Edison records on lateral ("needle") machines is answered by this memo dated February 1, 1923. "When a [Edison] dealer sells an Edison attachment to play Edison records on a needle machine he loses the future sale of an Edison instrument of some kind averaging $168.00 – yet he is sublimely ignorant of this and jobber as well." Never let it be said that the Great Man overestimated his sales force. *Courtesy of René Rondeau.* Value code: VR)

3-43. The original box and instructions for an "O-M-C" attachment, which is itself missing. Judging by the prominence given Operaphone records, it begins to be clear that the "O-M-C" denotes Operaphone Manufacturing Corporation, which produced the vertically recorded Operaphone disc of the late teens. *Courtesy of Charles McCarn*. (Value code: D complete attachment)

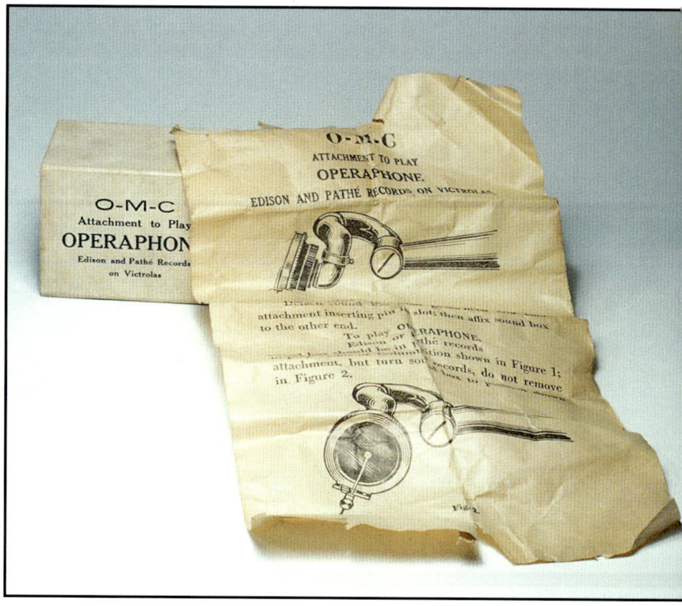

3-44. The Fletcher attachment was a popular device enabling Edison Diamond Disc Phonographs to play "all records." Selling for $8.50 in nickel plate, this attachment was not cheap by the standards of the time. However, once the Edison Phonograph owner realized the limited repertoire that was available on Diamond Disc records, he quickly reached for his wallet. *Courtesy of the Babcock House Museum, John H. Perschbacher.* (Value code: D)

3-45. The "Oro-Tone" was another device designed to make the Edison somewhat more "user-friendly," enabling its owner to play other makes of record. *Courtesy of Richard and Nancyann Brown.* (Value code: E)

3-46. The Edison Disc Phonograph Model "A-150" was available with an "oxidized bronze" finish on its metal parts. This style of decorating hardware enjoyed a vogue during the first decade of the twentieth century and into the early teens. The "Balance" lateral attachment for lateral discs shown here features a matching finish. *Courtesy of Jon and Jacqueline Kelm*. (Value code: E)

3-47. The "A.R.L." Attachment is a clever device which "makes the Victrola a universal machine," meaning that it can play either lateral or Pathé style vertical cut records. *Courtesy of the Charles Hummel collections.* (Value code: E)

3-48. Some attachments to allow lateral phonographs to play vertically recorded discs were quite simple. Shown are small cast elbows sold by Everybody's Talking Machine Company of Philadelphia for 50 cents each. The soundbox was removed from a Columbia Grafonola, the elbow inserted in the arm, and the soundbox inserted in the elbow. In this way, vertical cut alignment was achieved, and the reconfigured soundbox could even be rotated back into the "rest" position, a common feature of Grafonolas. (Value code: G)

3-49. To today's collector, the Victor "Orthophonic" soundbox is a mixed blessing: it offers the finest acoustic reproduction attainable by a major manufacturer, yet is also subject to failure due to the deterioration of the "white metal" alloy used in all but the earliest examples. A warehouse full of these would be a collector's dream – brand new, right out of the box and untouched by time… (Value code: E)

3-50. A Victrola "1-1," circa 1926, equipped with a device manufactured by the Home Recording Company of New York City. Not shown are the "blanks" on which the owner could record family, friends or amateur operatics. *Courtesy of the Domenic DiBernardo collection.* (Value code: D)

3-51. Perhaps the most elegant of the disc home recording devices was this British "Neophone" attachment. The nickel-plated mechanism included a feedscrew, which drove the recording head across specially prepared discs. Note the Bettini-like design of the recorder, having a "spider" with legs of differing lengths supporting the cutting stylus. *Courtesy of Sam Sheena.* (Value code: D)

3-52. One of the early advantages of disc versus cylinder record production was the ability to make metal stampers that would allow hundreds of copies to be produced from a single master recording. Shown is a 7" Berliner metal stamper or matrix of a recording made on May 29, 1898. This unnumbered matrix features an unissued speech on Freemasonry by an orator whose undecipherable signature appears on the disc. Of interest is the notation at the top, "Revolve 65 per minute." Such information was never overtly marked on Berliner discs themselves, even though the recitations of speeches might have been recorded at a speed slower than 78rpm during this early period. (Value code: VR)

3-53. A Canadian 7" Berliner record of unusual interest and historical significance. The record consists of a celluloid skin over a brown pasteboard base (pressed by The Celluloid Company, New York). The recording, dated December 8, 1900, is titled, "The Gramophone," and was spoken by Emile Berliner's younger son, Edgar. (Value code: VR)

3-54. Early disc record piracy was a problem for Berliner in the United States and a concern for him in Canada. Unscrupulous companies sometimes used commercial Berliner discs to create "anonymous" stampers from which unauthorized copies were produced. After moving his operation to Canada in 1900, Berliner devised this clever, albeit short-lived solution: lightly inscribing the famous trademark he shared with Victor directly among the grooves of the record. *Courtesy of Michael F. Khanchalian*. (Value code: VR)

As noted in the catalogues of several early talking machine companies, the heart of a phonograph is its reproducer or soundbox. The larger companies expended a great amount of time and money conducting experiments in pursuit of improved soundbox designs. Most of these experimental models have been lost or destroyed over time, but occasional examples can be encountered in private collections. The following nine photographs document, for the first time, a group of Victor experimental sound boxes which were literally rescued from a dumpster some half-century after their creation in Victor's Camden, New Jersey, research facilities. Some retain original identification tags bearing mysterious numbers, but little precise dating information is available. Judging by the designs and materials, the remarkable group of experimental sound boxes dates from a 20-year period following 1905.

3-55. The briefly-sold Victor "Improved Concert" sound box (1910) served as the basis for these experiments in various needle bar mounts and pivots. (Value code: VR)

3-56. Tests of diaphragm designs and materials based on the "Victrola No. 2." (Value code: VR)

3-57. An entirely new design to the left looks crude and clunky in comparison to the gold-plated "Exhibition" to the right, modified with an elongated needle bar and top-mounted pivot adjustment. (Value code: VR)

3-58. Various needle bar geometry was explored in these three Victor designs. (Value code: VR)

3-59. In the quest for more volume and improved low-requency response, Victor explored the possibility of two parate diaphragms in one sound box. The model on the right employs twin diaphragms set at pproximately a 45-egree angle to each other. The example n the left used two parallel offset diaphragms, and eparate needle bars with a single set of pivots. (Value code: VR)

3-60. A pair of experimental sound boxes further probing the twin-diaphragm design. The example to the left used a single set of pivots with a wide upper needle bar, drilled out to improve compliance. The model to the right employed two sets of pivots, the upper set linked to the lower needle bar. The mechanical energy loss with this design must have been significant. (Value code: VR)

3-61. An interesting evolutionary group of Victor artifacts. To the left, the "Victrola" sound box may be the progenitor of the well-known "Victrola No.2." At upper center is an otherwise typical "Victrola No.2" soundbox, but with a series of holes drilled into (but not through) its body. At lower center is another "No.2" with a crude bezel and needle bar variation. To the right and dated 10-27-23 is an otherwise unknown "Victrola No.3" soundbox, resembling an "Exhibition" with a pressed-in retaining ring. The first and last items described in this caption may be considered the "missing links" in a family tree that previously contained only a "No.2" and a "No.4." (Value code: VR)

3-62. Three differing examples of another previously unknown Victor soundbox: the "Victrola Special." T resemblance to a circa 1910 "Improved Concert" is obvious, and cou suggest a similar date, though the aluminum diaphragm might point t the 1920s. Because similar diaphrag were used by Mobley prior to 1910, dating these prototypes is particular difficult. (Value code: VR)

3-63. Victor "Orthophonic" experimentation is evident in these models, and fortunately, data survives on the identification tags. To the left, an early model labeled "Western Elect. Co.'s Style." Western Electric Company was assisting in the development of Victor electric recordings in 1925. At upper center is a brass "Orthophonic" sound box featuring the "#4 type of needle bar mounting." The "Victrola No.4" soundbox was released in 1926. At lower right is a unique experimental design, labeled "Orthophonic Sound Box," and dated 6/25/26. (Value code: VR)

3-64. Gadgets are sometimes found appended in tandem to disc sound boxes. In this instance, an intricately designed volume control, with knurled collars, shares a Victor Exhibition sound box with a convenient grasping post (affixed to one of the retaining screws in the soundbox frame). The volume control is inserted between the jointed tone arm and the soundbox of a conventional Victor talking machine. Note the miniature scale complete with pointer that indicates the volume setting. *Courtesy of Donald E.R. Poisson.* (Value code: E)

3-65. This "Sound-Regulator for Phonographs" was patented by Arthur H. Ryder on January 1, 1907 (No.840,089). At that time, Ryder intended the apparatus to be used in conjunction with external horn instruments (either disc or cylinder). In the patent specifications, the device was to be inserted between the soundbox, or reproducer, and the horn. The same function is achieved here by adding it to the tone arm of a Victrola. Ryder's invention consisted of "an apertured valve-plate cooperating with [a] fixed plate." The "valve-plate is provided with concentrically-arranged oblong slots forming ports or openings. The volume of vibratory sound may be regulated at all times by turning the valve plate." *Courtesy of Donald E.R. Poisson.*

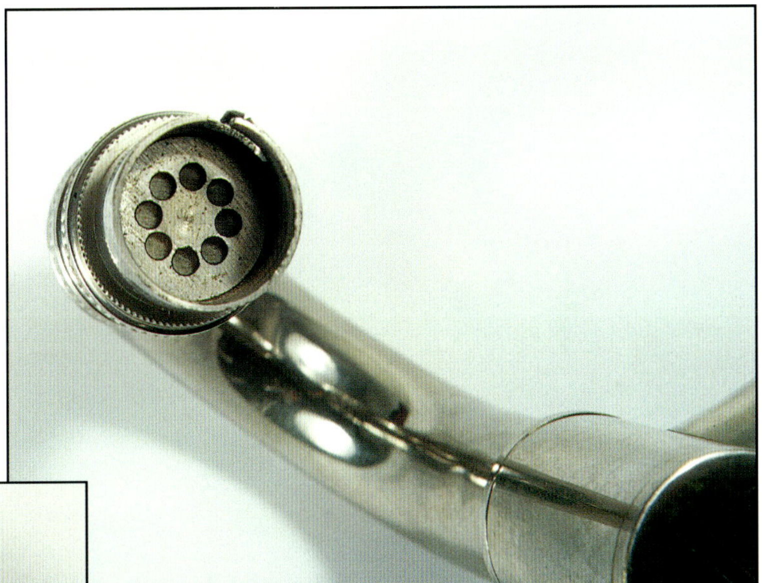

3-66. Another aftermarket volume control placed between the rubber flange and the body of a Victor "Exhibition" soundbox. *Courtesy of Pat and Jamie Jones*. (Value code: F)

3-67. A volume control fitted into a soundbox. The dampening ring mounted just above the diaphragm pivots at the top and is made stationary with a set screw. By gently adjusting this dampener (which has a rubber gasket mounted behind it) to make contact with the diaphragm, the reproduction may be softened considerably. *Courtesy of Steve and Ellie Saccente*. (Value code: E)

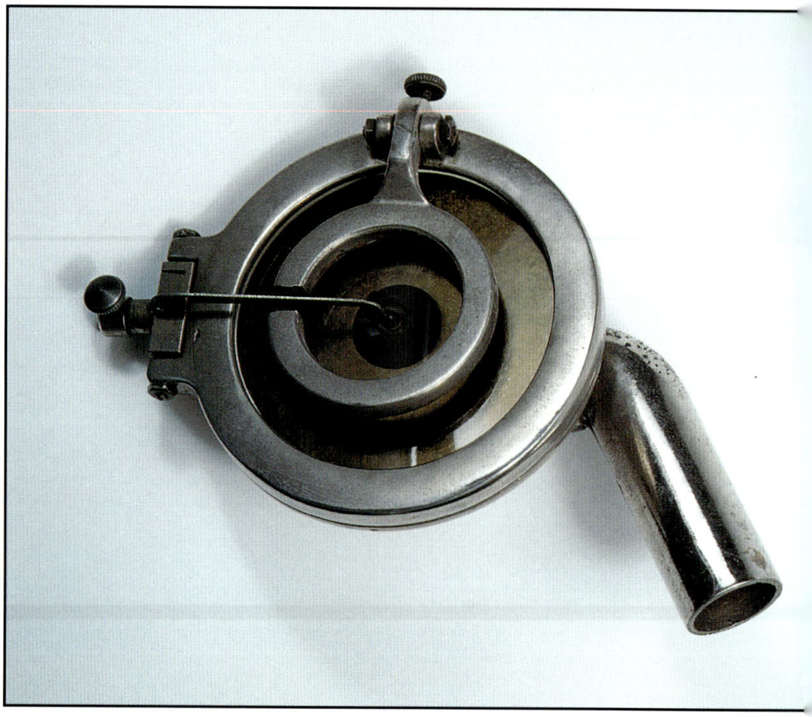

3-68. The "Velvet Tone" attachment for Victors and Victrolas was announced to th trade by A.D. Macauley of Columbia, Pennsylvania, in the February 1911 *Talkin Machine World.* A bulb-like weight counte balanced the soundbox, which rested on th shaped rod. According to an advertisemen "The 'Velvet Tone' Needle Balance is simi to a pair of weighing scales, as it lifts the excess weight of the sound box off the rec thereby permitting only a slight but sufficient pressure of the needle on the sound wave-lines of the record." *Courtesy of Steve and Ellie Saccente.* (Value code: E)

3-69. The "Mello-Tone Attachment" was first advertised in the March 1907 *Talking Machine World* as the "Universal Toner" ("It Opens and Shuts"). It was a device for modifying the tone (and volume) of a talking machine. "'The Universal Toner' is the only sound modifier made that is adapted to all styles and makes of Talking Machines or Phonographs, either cylinder or disc, and fits any size or shape of horn." Retail price was $1.00. Manufactured by the Mello-Tone Company of Springfield, Massachusetts. *Courtesy of Robin and Joan Rolfs.* (Value code: E)

3-70. This unmarked 2 7/8" diameter adjustable red rubber mute circa 1906 was designed to be inserted into the horn of a talking machine, exactly like the "Mello-Tone Attachment" it resembled. Newspapers and magazines from the twentieth century's first decade occasionally featured cartoons and letters to the editor complaining about noisy talking machines. The existence of muting devices such as this one attests to the reality of the problem, particularly in urban areas. *Courtesy of Yves Rouchaleau*. (Value code: E)

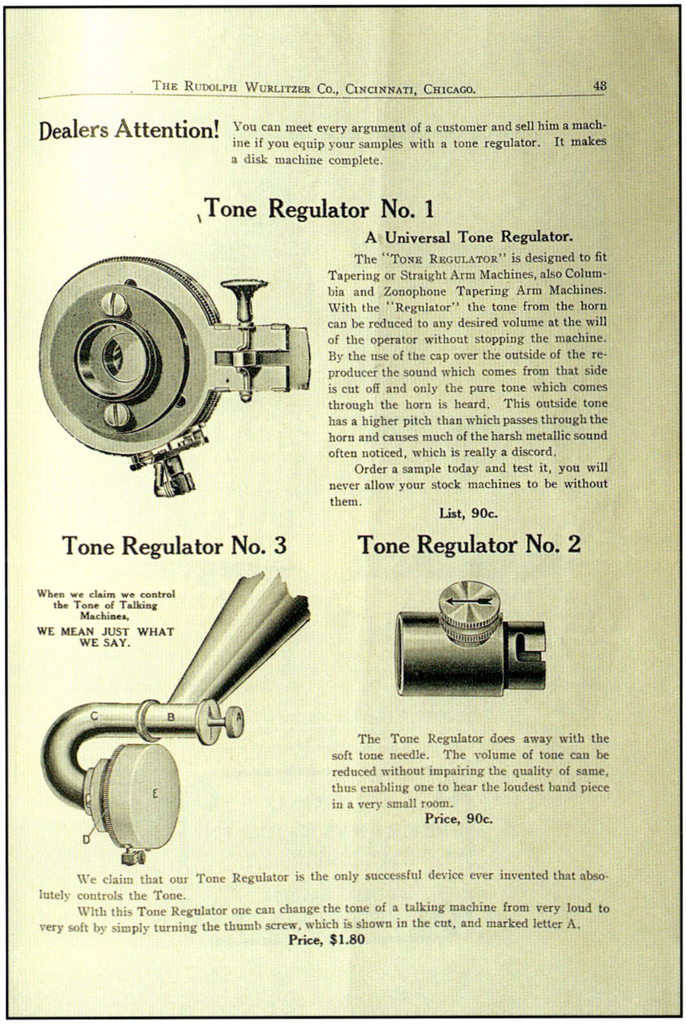

THE RUDOLPH WURLITZER CO., CINCINNATI, CHICAGO. 48

Dealers Attention! You can meet every argument of a customer and sell him a machine if you equip your samples with a tone regulator. It makes a disk machine complete.

Tone Regulator No. 1

A Universal Tone Regulator.

The "TONE REGULATOR" is designed to fit Tapering or Straight Arm Machines, also Columbia and Zonophone Tapering Arm Machines. With the "Regulator" the tone from the horn can be reduced to any desired volume at the will of the operator without stopping the machine. By the use of the cap over the outside of the reproducer the sound which comes from that side is cut off and only the pure tone which comes through the horn is heard. This outside tone has a higher pitch than which passes through the horn and causes much of the harsh metallic sound often noticed, which is really a discord.

Order a sample today and test it, you will never allow your stock machines to be without them.

List, 90c.

Tone Regulator No. 3

When we claim we control the Tone of Talking Machines, WE MEAN JUST WHAT WE SAY.

Tone Regulator No. 2

The Tone Regulator does away with the soft tone needle. The volume of tone can be reduced without impairing the quality of same, thus enabling one to hear the loudest band piece in a very small room.

Price, 90c.

We claim that our Tone Regulator is the only successful device ever invented that absolutely controls the Tone.

With this Tone Regulator one can change the tone of a talking machine from very loud to very soft by simply turning the thumb screw, which is shown in the cut, and marked letter A.

Price, $1.80

3-71. The 1907 Rudolph Wurlitzer Company catalogue illustrated three types of Tone Regulators (actually, volume controls) for disc machines. *Courtesy of the Gfell family collection.*

3-72. An unusual muting device as seen on an Edison Diamond Disc Phonograph "A-450." With the grille removed, a reverse L-shaped rod with a knurled knob at the top may be seen. Rotating the knob inserts or withdraws the mute from the horn. Note also the unusual key incorporating a lyre in its design. *Courtesy of the Sanfilippo collection*. (Value code: VR)

3-73. The French Aéropho not only looked peculiar, it incorporated some unusual features. Most obvious to th eye is the brass horn, descri as follows in the Aérophone catalogue, "This new horn its correct shape gives the b rendering of disc record reproduction." The horn wa also sold separately to equip other makes of talking machine. The reader has certainly noticed the doors the front of the cabinet, and this introduces the next illustration. *Courtesy of Phil Le Ray.* (Value code: A)

3-74. The Aérophone bracket contained important mechanical elements. As the catalogue explained, "This new arm constitutes the latest talking machine improvement carried out, it allows… First: To play the same disc *soft* or *loud.* [A volume control.]… Second: To play with an external or internal horn, or the two at once. [This is certainly a first in the phonograph field, not excepting the ones constructed in your uncle's garage.] … Third: To make subtle adjustments according to the whim of the user." *Courtesy of Philippe Le Ray.*

3-75. Invention inspires imitation, and so here is a French-made take-off on the Auxetophone: "Le Tonnerre" (the thunderclap). It will fit any Victor or Gramophone tone arm talking machine. The air hose attaches to the nozzle at the top. *Courtesy of Jalal and Charlotte Aro.* (Value code: VR)

3-76. The "Auxetophone," sold in Great Britain and the United States, was not the only disc reproducing apparatus to employ the principal of compressed air forced across a vibrating diaphragm, which was known to greatly augment volume. In France, the motion picture company Gaumont developed a system of "sound" films (the "Chronophone") accompanied by specially-recorded disc records. Amplification was by compressed air. Here is the Gaumont sound box in its carrying box, with a spout to which the air hose attached on the right. *Courtesy of Marie-Claude Stéger.* (Value code: VR)

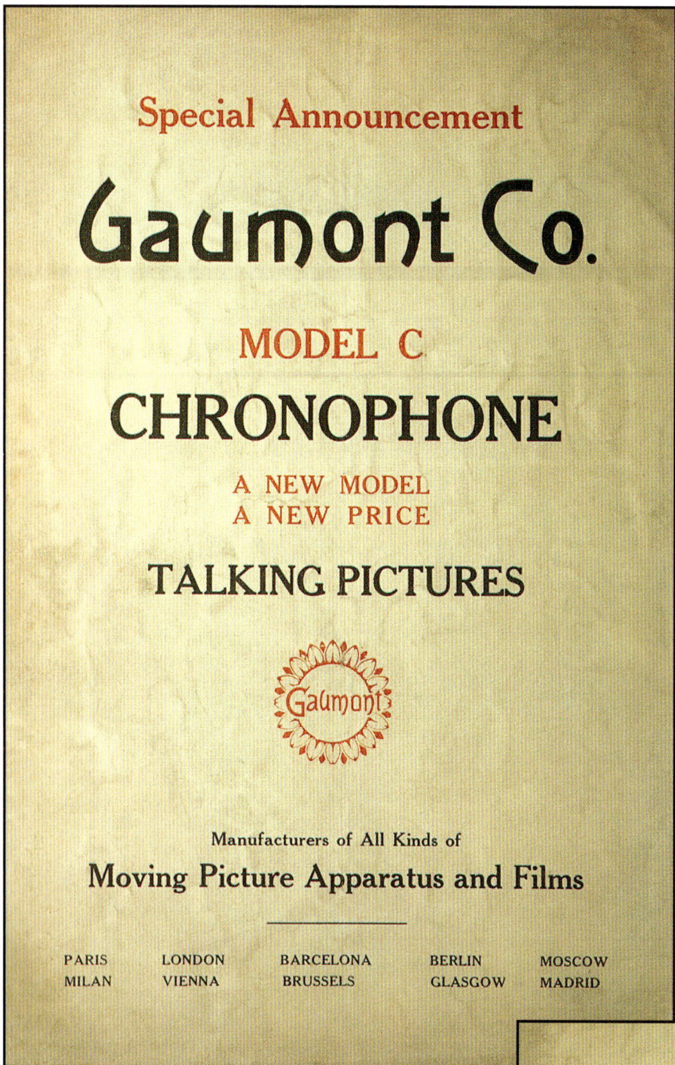

3-77.A Gaumont "Chronophone" catalogue, circa 1910.

3-78. The back cover of the Gaumont catalogue strikingly depicted the "Chronophone" in operation. The twin-horned phonograph employed the compressed air soundbox shown in illustration 3-76.

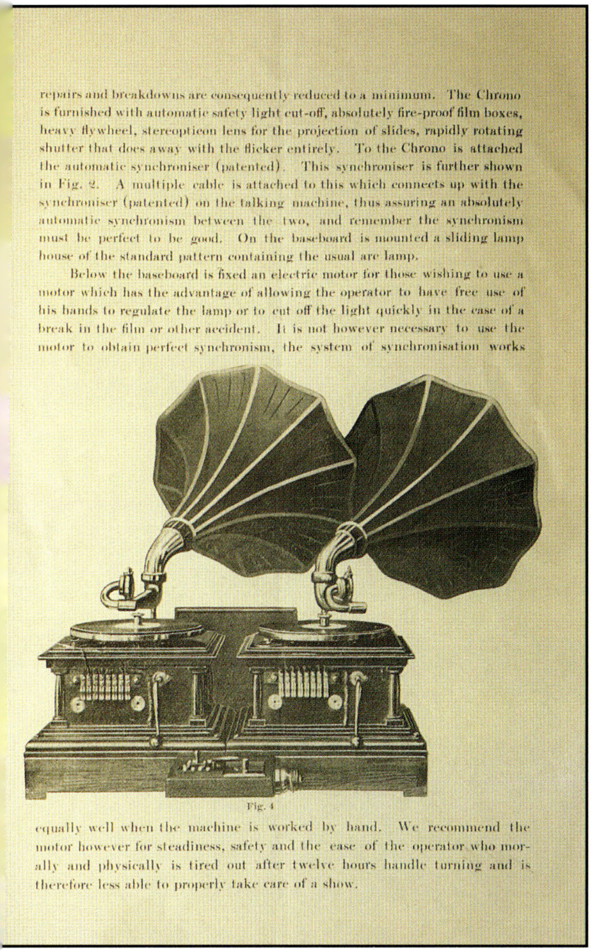
repairs and breakdowns are consequently reduced to a minimum. The Chrono is furnished with automatic safety light cut-off, absolutely fire-proof film boxes, heavy flywheel, stereopticon lens for the projection of slides, rapidly rotating shutter that does away with the flicker entirely. To the Chrono is attached the automatic synchroniser (patented). This synchroniser is further shown in Fig. 2. A multiple cable is attached to this which connects up with the synchroniser (patented) on the talking machine, thus assuring an absolutely automatic synchronism between the two, and remember the synchronism must be perfect to be good. On the baseboard is mounted a sliding lamp house of the standard pattern containing the usual arc lamp.

Below the baseboard is fixed an electric motor for those wishing to use a motor which has the advantage of allowing the operator to have free use of his hands to regulate the lamp or to cut off the light quickly in the case of a break in the film or other accident. It is not however necessary to use the motor to obtain perfect synchronism, the system of synchronisation works

Fig. 4

equally well when the machine is worked by hand. We recommend the motor however for steadiness, safety and the ease of the operator who morally and physically is tired out after twelve hours handle turning and is therefore less able to properly take care of a show.

3-79. The interior of the Gaumont catalogue illustrated a different "Chronophone" configuration, employing two modified Victor "IV" talking machines – dispensing with the complicated compressed air amplification.

3-80. The "Star Phonostop" (Patent Applied For) could be adapted to most any cabinet talking machine. Two screws held the device in place near the edge of the turntable. When the phonograph arm reached a pre-set point it tripped the rubber-tipped lever, causing a brake (far right, originally sheathed in rubber) to press against and stop the turntable. *Courtesy of the Scott and Denise Corbett collection.* (Value code: F)

3-81. Francis A. Nolan (or "Nolen" if the carton is to be believed) must have recognized how thoroughly annoying it was to adjust the automatic stop on an Edison Diamond Disc Phonograph. One needs to remove the turntable and use a screwdriver before the brake can be altered. Although the plan must have been to end all Diamond Disc records in the exact same spot, in reality the mechanism stops too soon, or doesn't stop at all. Enter the "Nolan Stop," which could be substituted for the factory-installed arm that tripped the brake lever. The device was easily set, with the help of the handy dial, to the stopping point of each disc. *Courtesy of the Scott and Denise Corbett collection.* (Value code: E)

3-82. This 3" celluloid disc labeled "Gramophone hints" suggested to owners that they leave it on their turntables... *Courtesy of Garry James*. (Value code: G)

3-83. The reverse side included an advertisement for British "Songster" needles and helpful tips on using and maintaining a talking machine. *Courtesy of Garry James*.

3-84. This accessory, the Pacent Phonovox, was an electric pick-up with which the owner of a Victrola or Grafonola could replace his mechanical soundbox, and play records electrically through a radio. (Value code: G)

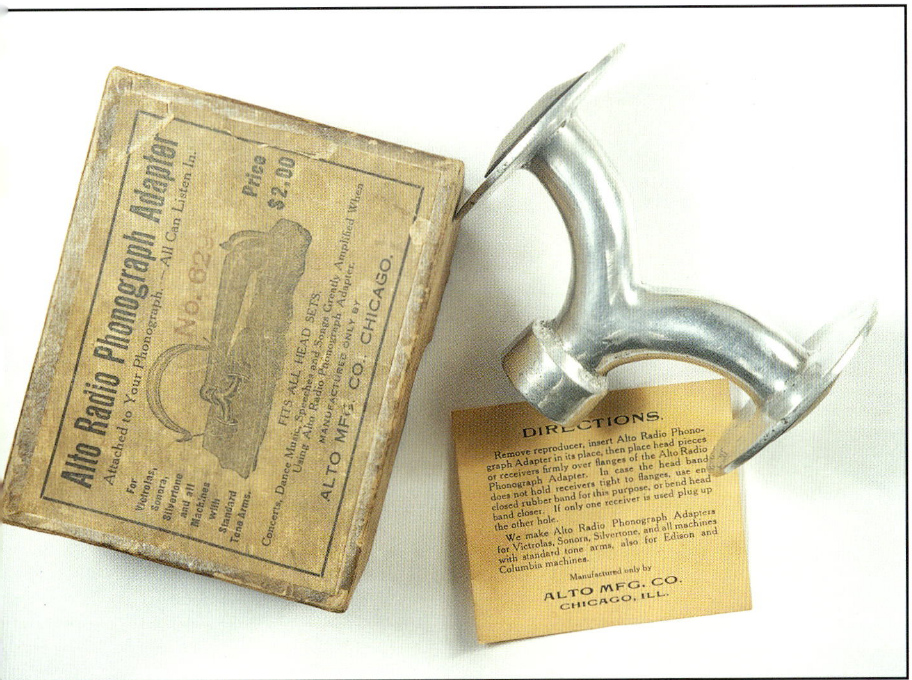

3-85. Early radios employed headsets rather than loudspeakers. In order for the entire family to listen to radio broadcasts, gadgets such as this "Alto Radio Phonograph Adapter" were developed in the early 1920s. In this way, talking machines briefly enabled radio to be heard by a wider audience. *Courtesy of Yves Rouchaleau*. (Value code: F)

3-86. The "Duplex Victrola Attachment" was a device similar to the "Alto," but with a clamp to secure the headphones. *Courtesy of the Scott and Denise Corbett collection*. (Value code: F)

3-87. The mid-1920s "Oro-Tone No. 100 Combined Recording, Reproducing and Radio Attachment" seemed to breathe new life into ten-year-old talking machines. This Edison Diamond Disc Phonograph from the teens is ready to record the family on special metallic discs, play them back, or allow the gang to listen to their radio through the phonograph horn rather than a headset. *Courtesy of the Scott and Denise Corbett collection*. (Value code: E)

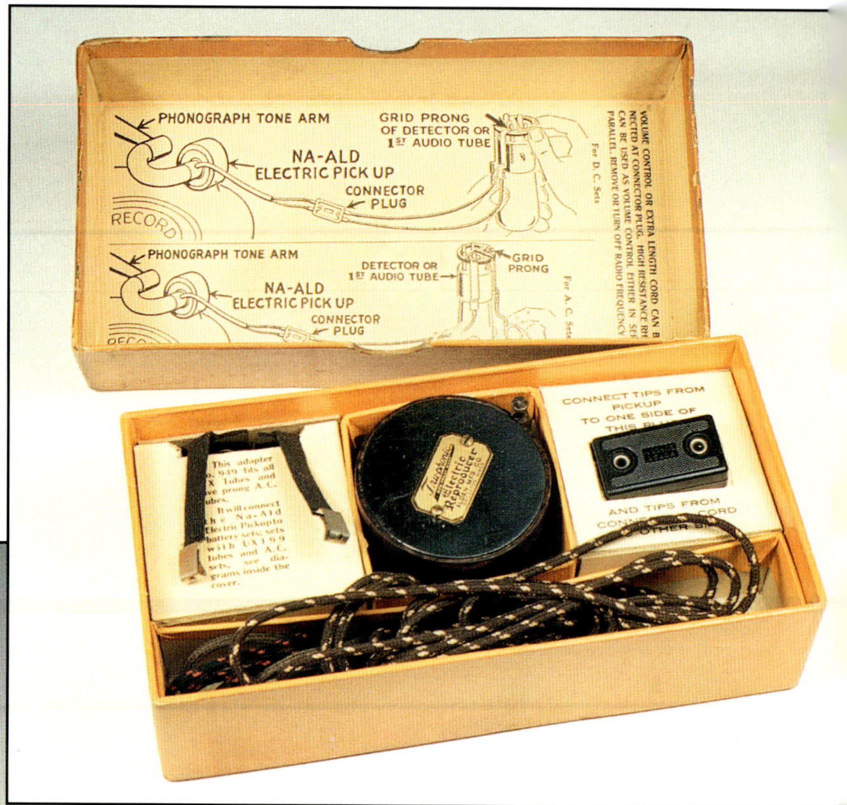

3-88. The "Tru-phonic" electric reproducer enabled the owner of an acoustic disc phonograph to play records electrically through an AC or DC circuit radio. This is model No. "50Z," originally priced at $5.00. Made by the Alden Manufacturing Company, Brockton, Massachusetts. *Courtesy of Dan and Sandy Krygier*. (Value code: F)

3-89. Historically, the "reproducer" was the sound-reproducing device found on cylinder phonographs, while a similarly functioning device on disc phonographs was called the "sound box." Exceptions exist – the most notable being Edison "Diamond Disc Reproducers." Shown is possibly the rarest exception. When the Columbia Phonograph Company was attempting to break into the burgeoning disc talking machine market in the late 1890s, the only way it could be done was to devise a "Disc Graphophone" which would closely resemble cylinder Graphophone technology. Shown is the "reproducer" of a Columbia "Toy" Disc Graphophone: virtually identical to that of a cylinder Graphophone. This disc incarnation exhibits a longer shank than those found on cylinder models, as well as a ball-shaped "universal joint." An ordinary cylinder reproducer box was used, with the "Disc" nomenclature rubber-stamped on the label. (Value code: VR)

3-90. An unusual accessory lamp for the interior of a cabinet phonograph. Standing approximately 4" high, this gold-plated accessory is unmarked except for the "PATENTS PENDING" on the top. *Courtesy of John D. Robles*. (Value code: F)

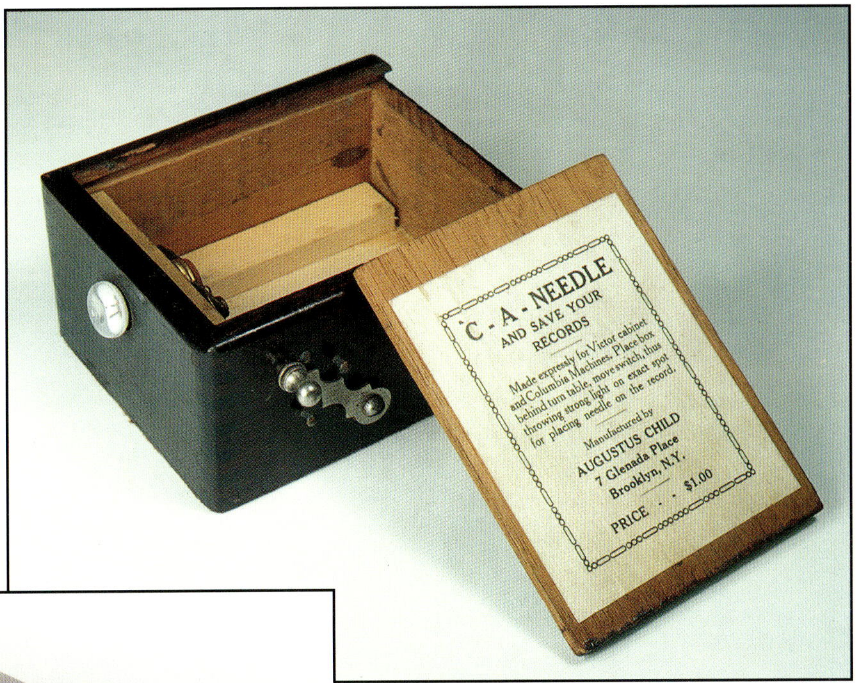

3-91. Of the comparatively few battery-operated accessory lights marketed for the dark recesses of Victrolas, Grafonolas, and other cabinet phonographs, this one is possibly the most aesthetic. Its dovetailed wooden box happens to harmonize with the machine. Unfortunately for the Brooklyn manufacturer, its trade name must rank among the clumsiest: "C-A-Needle." (Value code: F)

3-92. Another interior lamp - this one particularly suited for Edison Diamond Disc Phonographs, which used cast iron bedplates to support the mechanism. These machines discouraged owners from mounting anything on them – except this unmarked battery-operated accessory which was designed to be attached to vertical wooden surfaces within the cabinet, including the playing compartment or record storage areas. *Courtesy of Gus Schild*. (Value code: F)

3-93. A relatively large (5" square) tin of Edison grease "For Edison Diamond Disc and Amberola Phonographs." *Courtesy of the Scott and Denise Corbett collection*. (Value code: F)

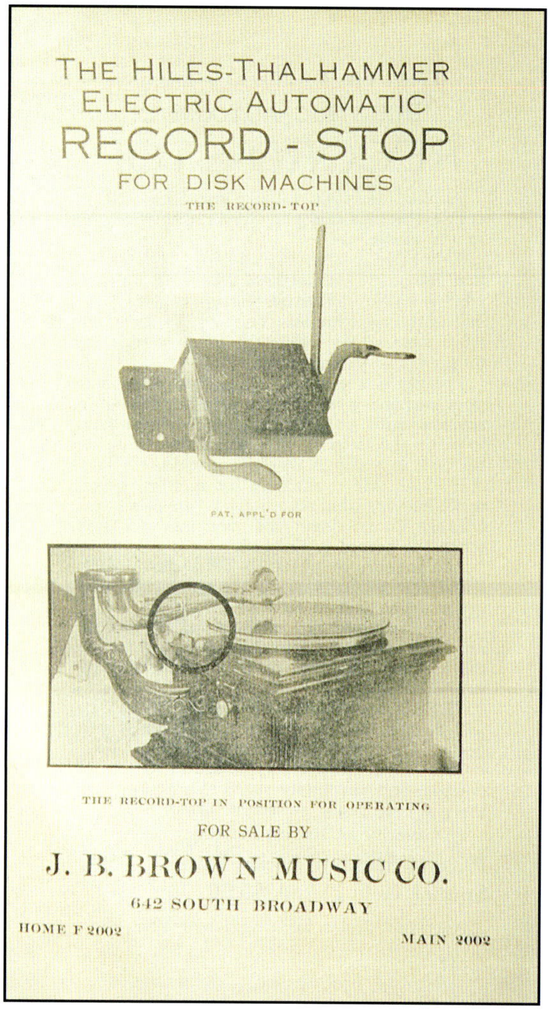

3-94. An advertising leaflet for the "Hiles-Thalhammer Record-Stop" showed the battery-operated device in place — ready to stop the turntable once the record was finished playing. It also demonstrated the need for careful proofreading. Note that "Stop" is misspelled "Top" in both illustration titles. Who was to blame before "spell-check?" *Courtesy of the Scott and Denise Corbett collection.* (Value code: H)

3-95. The attributes of fine spruce in the manufacture of violins and guitars are the stuff of legend. There are stories of master luthiers "sounding" pieces of cut spruce to determine which would perform to high standards. Mystery surrounds the violinmakers of Cremona and their fabled secret processes employed to treat the wood. In the 1920s, the prosaically named Diaphragm Company of Cleveland, Ohio, marketed a soundbox equipped with a "Violin Spruce" diaphragm, perhaps in hopes that the prestige of a rare Amati might be associated with its product. The scarcity of this soundbox suggests that the use of spruce as a diaphragm was more romantic than practical. *Courtesy of Tom Sideleau.* (Value code: F)

3-96. During the latter half of the twentieth century's first decade, Victor relentlessly pursued and eliminated small talking machine companies whose products could be prosecuted under United States patent No.534,543. Any manufacturer whose machine relied on the groove of the record to advance the needle of the soundbox could expect to hear from Victor's attorneys. One of these unfortunate firms was the Duplex Phonograph Company, whose twin-horned product appeared in 1905. By 1909, exhausted by Victor's litigation, the company made a last, forlorn attempt to escape Victor's patent by incorporating this "mechanical feed" into its machines. On this late example, a coiled spring gently pulled on a small arm projecting from the horn yoke, thus exerting a slight clockwise force, theoretically propelling the sound box toward the center of the record. The judge remained unconvinced, and Duplex production ceased by the end of 1910. To bring the story into sharp relief, we will quote from *Victor versus the Duplex Phonograph Company*, suit upon patent No.534,543 in United States Circuit Court, decision (in favor of complainants) of Judge Knappen, published May 27, 1909, "...the defendant began using, at least experimentally, near the bearing of the bracket which supports the framework carrying the horns... a weak, unadjustable, spiral spring, one end being attached to the bracket and the other end to a pin projecting from the framework carrying the horns... The defendant characterizes the pressure of this coil spring as an 'elastic mechanical feed'... I am not impressed with this contention... In my opinion, the defendant's machines are constructed and sold, and capable... of use both with and without the spring..." *Courtesy of Garry James.*

3-97.Another creative attempt to circumvent Victor's formidable United States patent No.534,543 was embodied within this 1910 Sonora phonograph with soundbox feed device at the rear of the playing compartment. *Courtesy of Brady Jeffcoat.* (Value code: VR)

3-98. Despite the impending expiration of patent No.534,543 (the "Berliner Patent"), Victor refused to throw in the towel, and attempted to eliminate Sonora. The Sonora utilized a true mechanical feed, but the courts decided it would not operate in the absence of a record. Now it was Sonora's turn not to give in. The independent firm went after patent No.534,543 itself, arguing that the legal Godzilla was invalid. The courts agreed, leaving Victor scrambling to appeal while rattling every saber in its arsenal at the competition. It was all rather academic by this point, the expiration date of the Berliner Patent, valid or not, was nigh. So, Sonora ebulliently introduced new models without the mechanical feed. *Courtesy of Brady Jeffcoat.*

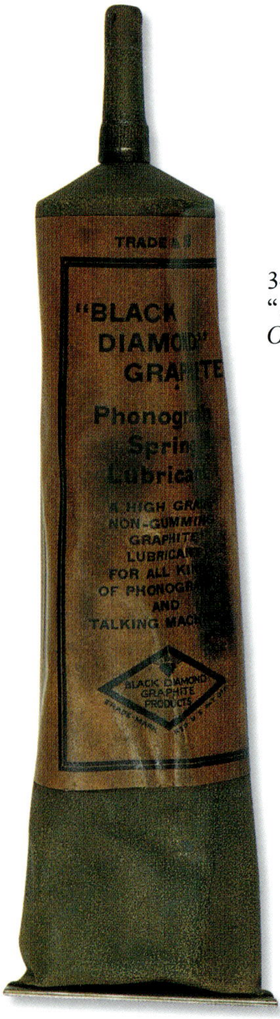

3-99. A tube of "Black Diamond" graphite lubricant, "…for all kinds of phonographs and talking machines." *Courtesy of Gus Schild.* (Value code: G)

3-100. A custom container whereby a Berliner "Improved" Gramophone could be disassembled and safely carried to a desired destination, along with a supply of records. *Courtesy of Gilbert (Bert) Pasley.* (Value code: VR)

3-101. This Compagnie Française du Gramophone machine (roughly equivalent to a Victor "II") sits atop its elegant maroon leatherette-covered carrying case. The horn, of course, required separate handling. (Value code: C, case only)

3-102. A carrying case with a surprising pink interior was used by the original owner to transport a Columbia "AJ" Disc Graphophone. *Courtesy of Walter and Carol Myers.* (Value code: C, case only)

3-103. A Victor "Monarch Special" could be packed into this case for destinations unknown. The horn, as usual, would travel separately. (Value code: C, case only)

3-104. Before the advent of true integrated "suitcase" portable phonographs in the teens and twenties, some manufacturers experimented with marketing machines in conveniently accessible carrying cases. This unmarked wooden phonograph sits in its velour-lined, leather covered carrier measuring 11 1/2" wide x 13 1/4" deep x 16 1/2" high. *Courtesy of Brady Jeffcoat*. (Value code: C, machine and case)

3-105. This General Phonograph Company Model "E" had the added advantage of being operable while sitting in its case. Measures 16 1/4" wide x 12 3/4" dee x 8 1/2" high. *Courtesy of Brady Jeffcoat*. (Value code: C machine and case)

3-106. Two Edison record dusters featuring embossed images of the great inventor flank a red velveteen record duster with an attached celluloid needle container. *Courtesy of Charles McCarn.* (Value code: G, F, G)

3-107. One of the phonograph dancing toys lesser known to collectors is the "Phono-Movies," which of course had nothing to do with films. Patented February 11, 1919, these toys were introduced in May 1919, as noted in the June issue of the *Talking Machine World*. The "complete attachment, with actors" retailed for $1.50. Manufactured by the Phono-Movie & Supply Company, 36 East 23rd Street, New York. *Courtesy of the Charles Hummel collections.* (Value code: C)

3-108. A French chocolate firm, Chocolat Ibled, arranged to have a small number of these disc machines and records manufactured with its own markings for advertising purposes. Note that even the sound box carries a milk chocolate colored label bearing the company name. The disc is vertically recorded, designed to be played with a sapphire ball, and there is little doubt that the machines and records were manufactured by Pathé. (Value code: C)

3-109. Pathé sold this disc record in Morocco, and both record and sleeve were suitably marked. (Value code: G)

3-110. The link between astronomy and talking machines is tenuous. Nonetheless, the appearance of the Halley Comet in 1910 made its fleeting mark on the popular culture of the day, including this European sound box. *Courtesy of Yves Rouchaleau.* (Value code: E)

3-111. The Italian Fonotipia disc catalogue contained fine examples of serious vocal and instrumental music. As this 27 1/2" x 39" poster suggested, the records achieved a worldwide reputation, especially after Columbia began distributing them in the United States. *Courtesy of Jalal and Charlotte Aro.* (Value code: C)

3-112. The phenomenal success of Victor's "His Master's Voice" trademark inspired numerous satires, some of which used alcohol as the comedic premise. This 9 1/4" x 13" lithographed "tin" sign proves that the recognition of Victor's logo could sell (with questionable legality) Kentucky rye whiskey as well as talking machines. The sign was manufactured by "Chas. W. Shonk Co. Litho., Chicago." *Courtesy of Charles McCarn.* (Value code: E)

3-113. Gianni Bettini moved his cylinder talking machine enterprise to Paris in 1902, and eventually began to design and market disc talking machines and records. This 6 7/8" diameter Bettini disc is a rare survivor of the period, distinguished by its modest size as well as its provenance. *Courtesy of Yves Rouchaleau*. (Value code: VR)

4. The most common form of "talking hine for hire" was coin-operated. The antages of an automatic mechanism are ous. However, in small venues such as a lunch nter or saloon, the machine could be placed on r and coins given directly to the proprietor. s Deutsche Grammophon machine, circa 1903, iteresting in several respects. It was sold in ope and Great Britain under various numeric gnations, although it was manufactured in the ted States by Victor. It is equipped with or's "rigid arm" which preceded the tapering e arm. A lid and sides have been added to the of the machine in order to confine the ambient nd. The most obvious addition is the set of d-held earpieces through which the patron ld hear the music. *Courtesy of the Philippe Le collection*. (Value code: VR)

3-115. The "Phonopostal" was made in France, and could record messages on specially prepared post cards. A reproducer was supplied for auditioning the card before it was sent or for playing back a card received from a friend. These machines were sold to the public, or could have been used in a store where postcards were sold. A 1906 advertisement by the Société anoyme des Phonocartes ("Phonocards, Incorporated") depicted a girl recording one of the firm's cards, known as "La Sonorine" ("the illustrated talking post card"), and her father listening to it. *Courtesy of Philippe Le Ray.*

3-116. This Pathégraphe was a French home study and recording device. The records it made were called Pathégrammes. The apparatus is really a 1920s refinement of the "Pathéposte," an earlier instrument that performed the same function. *Courtesy of Yves Rouchaleau.* (Value code: B)

3-117. These European needle boxes imitate Japanese lacquer. No such stylish wooden boxes appeared in the United States. *Courtesy of Wilfried Sator*. (Value code: E)

3-118. A record cleaner from the late teens with a prescient message. (Value code: F)

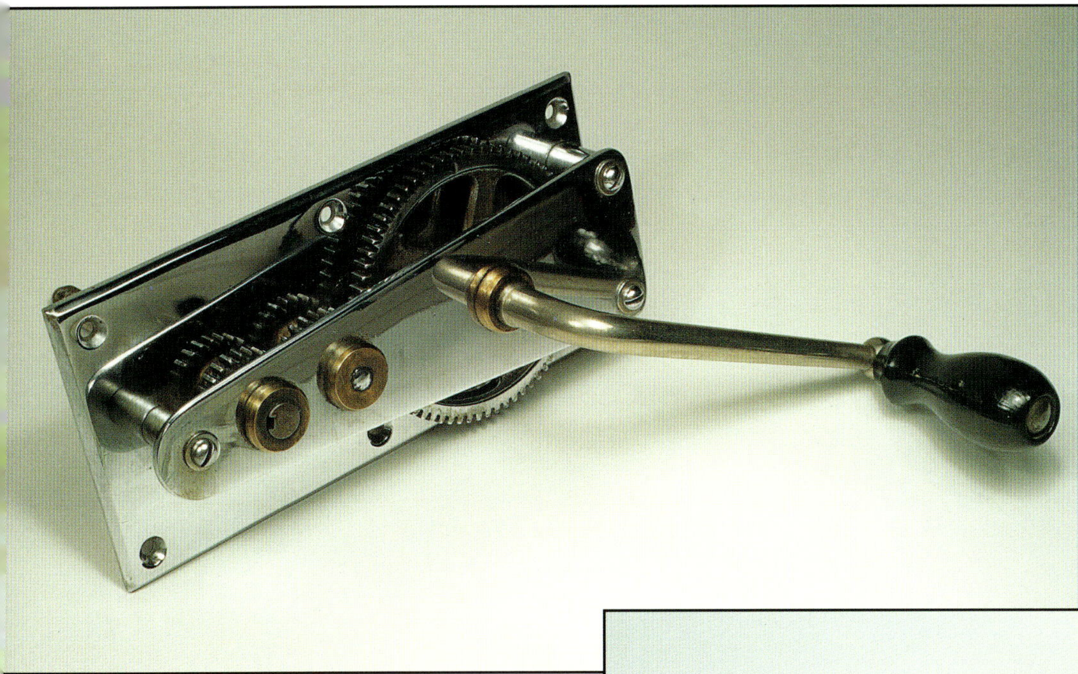

3-119.This substantial device allowed its owner to wind a Victrola in only one-third the normal time. A gadget of very high quality, likely one-of-a-kind, its three gears (riding in bronze bearings) are mounted on a massive nickel-plated brass plate measuring 6 3/4" x 3 1/4" and secured by a smaller triangular plate. Winding the crank (right) will rotate the winding shaft (left) at a ratio of approximately 3:1. The assembly is unmarked, and was originally found attached to a Victrola "XI" of the 1920 period. *Courtesy of Dan Reed*. (Value code: VR)

3-120. Small wooden needle boxes (the largest is 1 5/8" in diameter) from the 1905-1914 period. The descriptor "scratchless" and the direction to "use each six times" were meant to suggest a quality that, in all likelihood, these needles lacked. *Courtesy of Charles McCarn*. (Value code: G each)

3-121. The QRS Company was a well-known manufacturer of player piano rolls, and during the mid-teens it offered accompaniments to certain Victor records. In this way, a customer owning both a player piano and a phonograph could operate them simultaneously, to presumed paroxysms of musical ecstasy. The words printed on the piano rolls were in English and a foreign language appropriate to the selection. The catalogue shown is from 1918. *Courtesy of Alan H. Mueller*. (Value code: G)

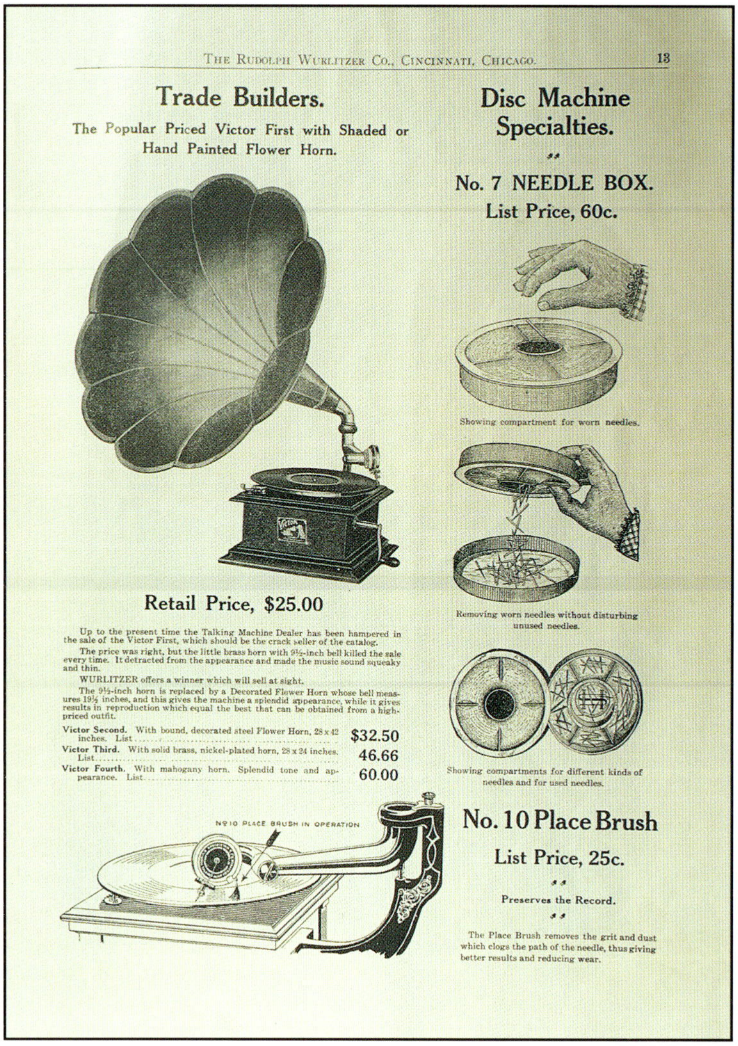

The Rudolph Wurlitzer Co., Cincinnati, Chicago. 13

Trade Builders.

The Popular Priced Victor First with Shaded or Hand Painted Flower Horn.

Retail Price, $25.00

Up to the present time the Talking Machine Dealer has been hampered in the sale of the Victor First, which should be the crack seller of the catalog.

The price was right, but the little brass horn with 9½-inch bell killed the sale every time. It detracted from the appearance and made the music sound squeaky and thin.

WURLITZER offers a winner which will sell at sight.

The 9½-inch horn is replaced by a Decorated Flower Horn whose bell measures 19½ inches, and this gives the machine a splendid appearance, while it gives results in reproduction which equal the best that can be obtained from a high-priced outfit.

Victor Second. With bound, decorated steel Flower Horn, 28 x 42 inches. List	$32.50
Victor Third. With solid brass, nickel-plated horn, 28 x 24 inches. List	46.66
Victor Fourth. With mahogany horn. Splendid tone and appearance. List	60.00

Nº 10 PLACE BRUSH IN OPERATION

Disc Machine Specialties.

No. 7 NEEDLE BOX.

List Price, 60c.

Showing compartment for worn needles.

Removing worn needles without disturbing unused needles.

Showing compartments for different kinds of needles and for used needles.

No. 10 Place Brush

List Price, 25c.

Preserves the Record.

The Place Brush removes the grit and dust which clogs the path of the needle, thus giving better results and reducing wear.

3-122. This 1907 Rudolph Wurlitzer Company catalogue page showed accessory horns for Victors, "Place" record brushes, and the seldom-seen Hawthorne & Sheble needle tray. *Courtesy of the Gfell family collection*.

3-123. An actual Hawthorne & Sheble needle tray, 4" in diameter. Compartments are marked, "HIGH GRADE," "SCRATCHLESS," "MUSICAL TONE," and "SYMPHONY" to denote various "tones" of needles. *Courtesy of Richard and Nancyann Brown*. (Value code: VR)

3-124. A framed canvas hanger showing the finishes available on Victrola cabinets, circa 1919. Note that there were eight different oak finishes, a number of which, in instruments over 80 years old, have begun to look nearly the same. 20 1/4" x 10 1/4". *Courtesy of Charles McCarn*. (Value code: D)

3-125. An assortment of keys for cabinet model phonographs. From left to right: early Brunswick, Edison dolphin, common Victor, Columbia, later Edison, later Brunswick. *Courtesy of Todd Emery*. (Value code: F each)

3-126. This Victor key is a departure from the usual types as seen in the previous illustration. *Courtesy of Fred Guenther.* (Value code: VR)

3-127. Before lockable Victrolas were supplied with the more familiar keys bearing a "V" within an oval, the Victor Talking Machine Company had employed more elaborate designs. Shown is an early Victrola key in an art nouveau style. *Courtesy of Charles McCarn.* (Value code: VR)

3-128. Empty phonograph packing crates are not universally appreciated by collectors, but to find a phonograph still housed in its original shipping container is something of an oddity. Shown is a Victrola "50" which was shipped in the early 1920s to Cressey & Allen, 534 Congress Street, Portland, Maine. *Courtesy of Fred Guenther*. (Value code: VR)

3-129. The Meyers Manufacturing Company of Chicag offered the "Univer- sal Record Recutter to an unwitting public, promising that it, "...can be attached to the reproducer of any talking machine on the market, and wi owing to its scientifical [sic] construction, remal the indentations th produce the music." No doubt this devic would "remake the indentations," but how much of the music would remai *Courtesy of the Babcock House Museum, John H. Perschbacher.* (Value code: E)

3-130. Witte's "4-in-1" Moviescope was really nothing more than an old-fashioned zoetrope which could be placed on the turntable of a phonograph. Manufactured by G.W. Witte, 1727 Ludlow Street, Philadelphia, Pennsylvania. Two new "picture films" (actually paper strips) were released every week, and the two could be purchased for five cents. *Courtesy of the Gfell family collection.* (Value code: B)

3-131. These three celluloid record dusters can't help but provoke lightheartedness. The yellow duster to the left is deliberately funny. "Hamburger" is probably the owner's name in the center duster, but customers surely had some fun with it. The duster to the right copied the famous biscuit moniker: "Uneeda." *Courtesy of the Scott and Denise Corbett collection.* (Value code: F each)

3-132. Peter Bacigalupi was a well-known West Coast distributor and retailer of talking machines. Based in San Francisco, Bacigalupi's empire was built upon the sale of Columbia Graphophones, Edison Phonographs, and (later) Victor talking machines, along with projecting machines, films, and a variety of related paraphernalia. The cardboard box shown here contains a dozen Edison cylinder records sold by Bacigalupi from his 933 Market Street address. This building served as his headquarters from November 1898 until its destruction in the San Francisco earthquake and fire of April 18, 1906. More clues to the exact age of these artifacts become evident through examination of the attractive box label, which lists prices of various Edison Phonographs, thus dating the box from the period between November 1899 to May 1900. *Courtesy of Brice D. Paris.*

3-133. "The Hughes Needle Starter, Manufactured by the Hu-Stylus Corp., Philadelphia, Pat. Pending." The two small thumbnuts on the right adjust the lateral movement of the forked lever. By installing this device alongside a phonograph's tone arm, and adjusting the nuts properly, the fork may be snapped into position for either 10" or 12" discs. (Value code: E)

3-134. Arthur J. O'Neill was a man of many parts: an amateur poet, traveling salesman, entrepreneur, and inventor, among other things. As more and more scrutiny has been directed at his life and considerable *oeuvre*, the greatness of his "master plan" to solve the Chicago talking machine muddle has become increasingly clear. In turn-of-the-twentieth-century Chicago, talking machine companies existed that sold disc instruments and records having non-standard spindle sizes. The Standard Talking Machine Company employed a spindle of approximately 1/2" diameter. The Great Northern Manufacturing Company sold "Harmony" machines, which used a 3/4" spindle. In 1907, O'Neill introduced the Aretino machine. It played doughnut-like records that had a 3" hole in their middles. At the heart of this enterprise was United States patent No.874,985, granted to O'Neill on December 31, 1907 (Aretino artifacts frequently reflect an incorrect patent date, as seen here). A reading of the patent reveals that O'Neill intended his Aretino record to be the resolution of the spindle wars, adaptable to any disc talking machine on the planet through the use of adapter rings. Shown here is one of the rarest of these adapters, made to fit the turntable of a Standard Talking Machine. *Courtesy of Kurt R. Nauk III.* (Value code: F)

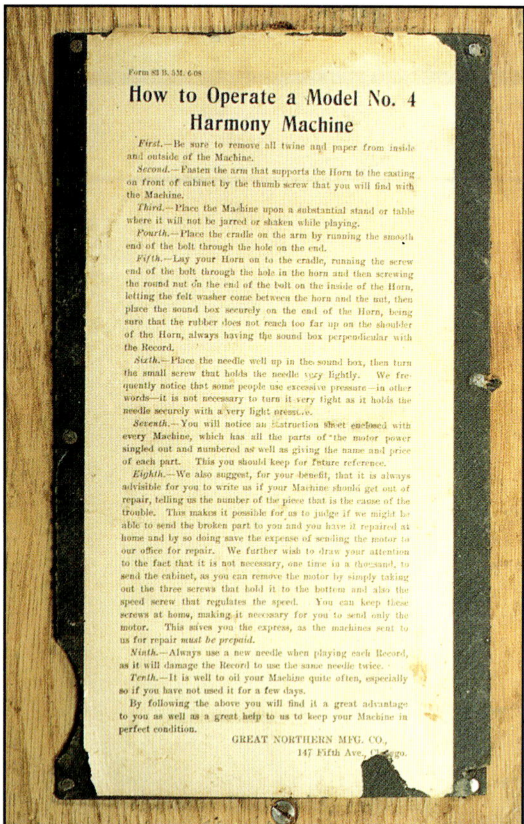
Form 83 B. 5M. 6-08

How to Operate a Model No. 4 Harmony Machine

First.—Be sure to remove all twine and paper from inside and outside of the Machine.

Second.—Fasten the arm that supports the Horn to the casting on front of cabinet by the thumb screw that you will find with the Machine.

Third.—Place the Machine upon a substantial stand or table where it will not be jarred or shaken while playing.

Fourth.—Place the cradle on the arm by running the smooth end of the bolt through the hole on the end.

Fifth.—Lay your Horn on to the cradle, running the screw end of the bolt through the hole in the horn and then screwing the round nut on the end of the bolt on the inside of the Horn, letting the felt washer come between the horn and the nut, then place the sound box securely on the end of the Horn, being sure that the rubber does not reach too far up on the shoulder of the Horn, always having the sound box perpendicular with the Record.

Sixth.—Place the needle well up in the sound box, then turn the small screw that holds the needle very lightly. We frequently notice that some people use excessive pressure—in other words—it is not necessary to turn it very tight as it holds the needle securely with a very light pressure.

Seventh.—You will notice an instruction sheet enclosed with every Machine, which has all the parts of the motor power singled out and numbered as well as giving the name and price of each part. This you should keep for future reference.

Eighth.—We also suggest, for your benefit, that it is always advisible for you to write us if your Machine should get out of repair, telling us the number of the piece that is the cause of the trouble. This makes it possible for us to judge if we might be able to send the broken part to you and you have it repaired at home and by so doing save the expense of sending the motor to our office for repair. We further wish to draw your attention to the fact that it is not necessary, one time in a thousand, to send the cabinet, as you can remove the motor by simply taking out the three screws that hold it to the bottom and also the speed screw that regulates the speed. You can keep these screws at home, making it necessary for you to send only the motor. This saves you the express, as the machines sent to us for repair *must be prepaid.*

Ninth.—Always use a new needle when playing each Record, as it will damage the Record to use the same needle twice.

Tenth.—It is well to oil your Machine quite often, especially so if you have not used it for a few days.

By following the above you will find it a great advantage to you as well as a great help to us to keep your Machine in perfect condition.

GREAT NORTHERN MFG. CO.,
147 Fifth Ave., Chicago.

3-135. The Chicago-based Great Northern Manufacturing Company was associated with a number of talking machine brands whose common denominator was the ability to play only those records specially made for them. This "Harmony" machine bears unusually complete instructions on the bottom of its cabinet. The instrument itself was constructed by the Hawthorne & Sheble Manufacturing Company.

3-136. Still more directions could be found beneath the turntable of the "Harmony" No. "4." As can be seen, access to the motor was direct, since the cabinet did not fully enclose it.

3-137. This clever contraption is a "universal" bracket and arm for adapting practically any "front-mounted" disc talking machine to "back-mounted" operation. The earliest configuration for disc instruments involved a horn projected straight out the "front" of the machine, supported by a set of extension arms. A design improved in the use of space began to be adopted by most companies around 1905: the horn was mounted over the top of the machine, which required the use of a separate "tone arm" to convey the sound, and a rigid bracket to support the elements. The attachment shown here mimics the conventional appearance of the later style equipment: a decorated cast iron bracket, "tapered" tone arm and swiveling "elbow" to secure the horn. What makes this design unique, however, is an eccentric joint at the top of the bracket to which the arm and horn elbow are mounted. By manipulating this joint, the arm can be effectively lengthened or shortened. This would allow the device to be substituted for the "front-mounted" equipment of any number of talking machines, regardless of the size of their cabinets. We do not know who manufactured this interesting gadget, but it was probably done without the imprimatur of the Victor Talking Machine Company, whose distinctive horn elbow fitting and patented tapering arm were copied. (Value code: C)

3-138. During the United States' involvement in the "Great War" of 1914-1918, Victor produced this set of three records and two booklets titled, "First Aid French for American Soldiers." Hoping, or perhaps sensing that America's direct participation would be limited to less than two years, Victor cleverly pointed out that the set would be useful to "the soldier, the tourist or the business man…" *Courtesy of Nina and Dave Heitz.* (Value code: F)

3-139. Few phonographs can match the opulence of the Columbia "Grand" of 1912, a Grafonola talking machine disguised as a scaled-down grand piano. The interior appointments were lustrous with gold plating, and luxurious with green velour. Along with the expected record index, a drawer held tools, an oiler, needle sharpener, and sound box in plush splendor. *Courtesy of Doug Anderson.*

3-140. A French needle box "for Zonophone" illustrating the "Concert Grand." (Value code: D)

3-141. The interior features receptacles for new (left) and used needles (right).

3-142. Although slightly beyond the historical scope of this book, it's interesting to compare the appearance of this 1932 electric home recording outfit from an RCA "RAE-84" with earlier acoustic contraptions shown previously. Included is a small rubber retaining washer, an extra weight (curved) for the pickup, microphone, recording blank, and extra recording styli. *Courtesy of Greg Bogantz*. (Value code: F)

3-143. An HMV "Instantaneous Speed Tester" in original box. *Courtesy of Freddie and Jeanette Blair.* (Value code: F)

3-144. Two record dusters, the one to the left is the "Standard" Record Cleaner (50 cents) featuring an "oxidized bronze" finish such as was briefly supplied on Edison Disc Phonograph Model "A-150," prior to 1915. "Standard" and "Simplex" cleaners were made by Kirkman Engineering Corporation of New York City (originally the Standard Gramophone Appliance Company). The duster to the right can be found with various dealer advertising. (Value code: G)

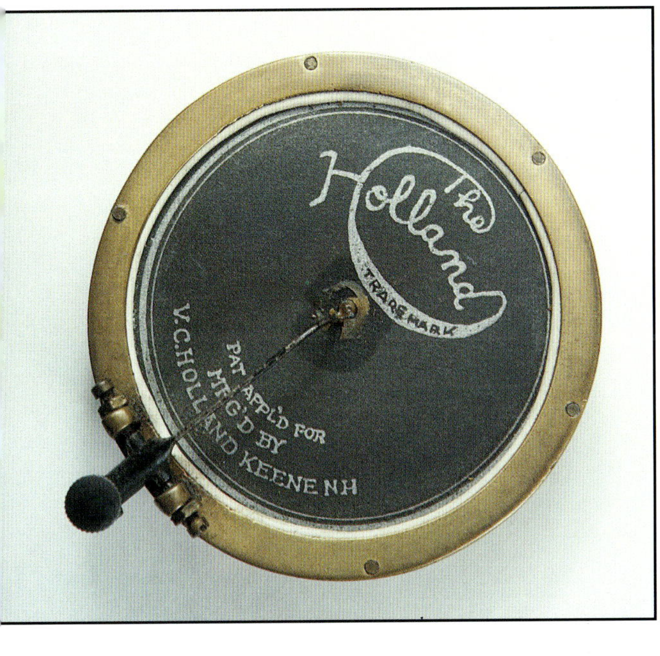

3-145. This 3" diameter "Holland" soundbox was manufactured (no doubt in small numbers) in Keene, New Hampshire. The fanciful trademark has a charmingly amateurish quality. *Courtesy of R. Chase.* (Value code: G)

3-146. An early "audiophile" soundbox designed to fit a Victrola. The device is heavy, made like a technical instrument, and fits into a custom wooden box with the maker's card: "Curtis A. Smith, Inventor of Talking Machine Attachments." 2 5/8" diameter. *Courtesy of Walter and Carol Myers.* (Value code: F)

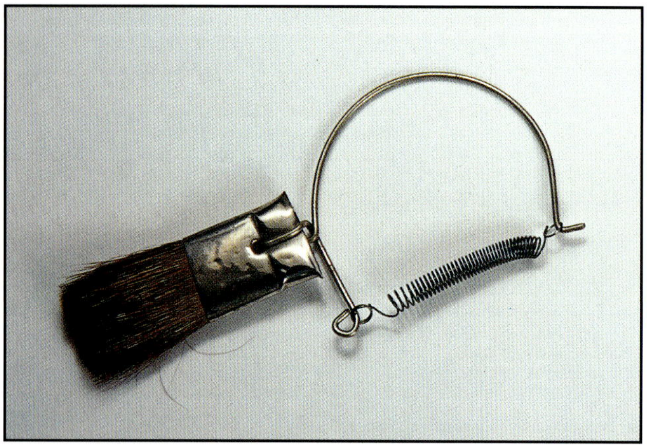

3-147. A spring-held record brush designed to fit a Victor "Exhibition" sound box, 2 1/2" long. *Courtesy of R. Chase.* (Value code: G)

3-148. When Emile Berliner moved his operation to Canada in 1900, his talking machine mainstay continued to be (as it was for Eldridge Johnson, his former contractor) the venerable "Improved Gramophone." It was a resourceful design, but was limited to playing only 7" diameter discs. Johnson subsequently introduced new models such as the Type "C" (1900) and the "Monarch" (1901) which were capable of playing the new 10" diameter discs (introduced in January 1901). Berliner's Canadian enterprise may have lacked Johnson's degree of innovation, but ingenuity was in plentiful supply, as demonstrated by the Canadian "Improved Gramophone" shown here. The availability of 10" discs and the public demand for them threatened to relegate the last of the "Improved Gramophones" to a quick obsolescence. Berliner, however, cleverly modified the brake, sound box rest, and speed control to accommodate the larger records. Unfortunately, nothing could be done about the positioning of the crank, which obligated the buyer of 10" discs to remove it each time a selection was played (as seen in the illustration). These were partially effective stopgap modifications meant to clear stocks and to add a few years of usefulness to this famous Gramophone, or to make previous customers happy by converting their instruments at the factory. The improved sound box dates from around 1904. (Value code: VR)

3-149. This phonograph sold by Newby & Evans in the late teens has two interesting features: both battery-operated. The removable auto-stop mounted on the spindle could be adjusted to point at the inner groove of the recording. When the steel needle touched the pointer, a battery circuit was closed, tripping a spring-loaded turntable stop. Also, the battery-powered lamp was turned off simply by closing the lid (note the switch at the extreme right). Opening the lid would automatically turn on the lamp. *Courtesy of the late Sam Saccente.* (Value code: E each)

3-150. A 2 3/8" diameter advertising mirror from a firm in Dusseldorf, Germany. *Courtesy of Wilfried Sator.* (Value code: E)

3-151. While most of the talking machine industry of the 1920s was focused on a "modern" image, this European soundbox managed to incorporate classical elements into its design. *Courtesy of Wilfried Sator*. (Value code: F)

3-152. Who says collecting phonographs is a gamble? This "Rouzette" game (copyright 1922) converted the parlor into a casino, and the phonograph turntable became a roulette wheel. High rollers shouting their bets; grandpa dropping the marble; even the "kiddies" going wild – what *IS* the world coming to? *Courtesy of Bill Feiner*. (Value code: D)

3-153. Edison included a small supply of grease with many of its Phonographs during the teens and twenties. Shown is a representative sampling of Edison grease jars. *Courtesy of Mark Trice*. (Value code: G each)

3-154. The Reproducto Manufacturing Corporation of Newport News, Virginia supplied wax disc blanks such as this with its home recording outfit. The company claimed that the disc could be shaved and recorded 50 times. *Courtesy of Bill Feiner*. (Value code: G)

3-155. Talking machine dealers were plagued by minor damage to phonograph cabinets in transit, in the store, or during delivery to customers' homes. Columbia offered its dealers materials and advice on "How To Repair Damage To Varnished Surfaces." Stains of various colors were provided. *Courtesy of Robin and Joan Rolfs*. (Value code: D)

3-156. A Brunswick needle display from the mid-1920s. Measures 15 1/2" wide x 9 3/8" deep. *Courtesy of Robin and Joan Rolfs*. (Value code: B)

3-157. The rear of the Brunswick needle display featured a drawer for storing the different types of needles offered. *Courtesy of Robin and Joan Rolfs*.

3-158. This 21" long Victor black flower horn was never unpacked from its crate. It is possible that the original owner purchased a wooden or more colorful aftermarket horn for his Victor "III" (or similar model) and placed this unopened crate in the attic, where it was forgotten. Distributed through the Talking Machine Company of Chicago, this horn was sold by the Voss Piano Company of Appleton, Wisconsin. *Courtesy of Robin and Joan Rolfs*. (Value code: VR)

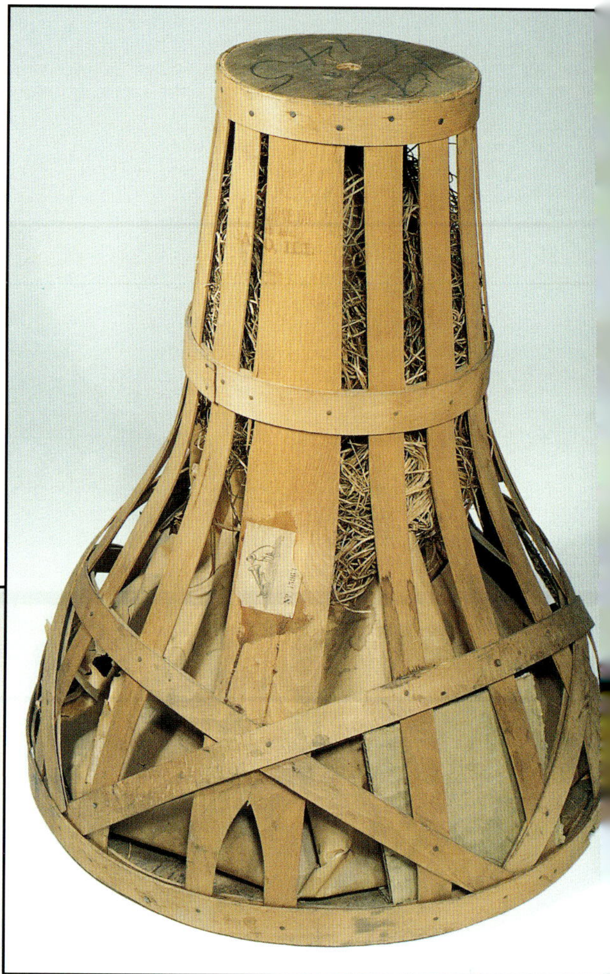

3-159. The "Gold Seal Phonograph Repeater" was a popular and reasonably efficient device that made the record repeat for as long as the spring motor would play. Available in various shades of celluloid including green and brown, the "Gold Seal" could not be adjusted for the length of the tune, and during play the contraption was rubbed by the record surface revolving beneath it. Despite longer tunes being abruptly curtailed, and the hiss of the gliding celluloid, these repeaters sold well. (Value code: G)

3-160. The reverse of the "Gold Seal Phonograph Repeater" packaging showed happy owners listening to repeated renditions at dinner, while cleaning the house, during a card game, and finally, during a well-deserved rest.

3-161. The "Encore Record Re-Player" was a clever contraption that employed a mechanical screw on top of a "gliding" steel frame. In use, records would be played by the soundbox needle while the red celluloid strip gently pushed against it, preventing the Encore from spinning around. As the needle approached the center of the record, the adjustable red celluloid strip would disengage from the needle, allowing the Encore to rotate 360 degrees and scoop the needle from behind, into the metal chute and onto the mechanical screw. The weight of the soundbox would fully engage the small black rubber wheel on the left side of the screw's shaft, thus driving the soundbox back toward the outer edge of the record. Finally, the soundbox would fall off the end of the screw, and the needle would slide off the metal chute and re-engage the red celluloid strip. *Courtesy of Dan and Sandy Krygier.* (Value code: F)

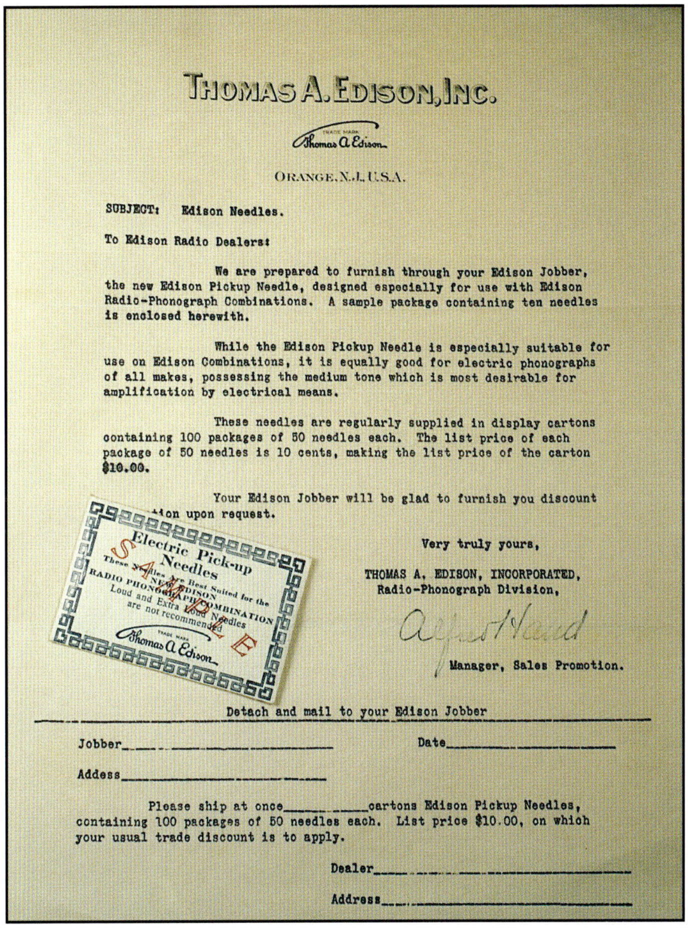

THOMAS A. EDISON, INC.

TRADE MARK Thomas A. Edison

ORANGE, N.J., U.S.A.

SUBJECT: Edison Needles.

To Edison Radio Dealers:

We are prepared to furnish through your Edison Jobber, the new Edison Pickup Needle, designed especially for use with Edison Radio-Phonograph Combinations. A sample package containing ten needles is enclosed herewith.

While the Edison Pickup Needle is especially suitable for use on Edison Combinations, it is equally good for electric phonographs of all makes, possessing the medium tone which is most desirable for amplification by electrical means.

These needles are regularly supplied in display cartons containing 100 packages of 50 needles each. The list price of each package of 50 needles is 10 cents, making the list price of the carton $10.00.

Your Edison Jobber will be glad to furnish you discount ...tion upon request.

Very truly yours,

THOMAS A. EDISON, INCORPORATED,
Radio-Phonograph Division,

Alfred Hand

Manager, Sales Promotion.

Detach and mail to your Edison Jobber

Jobber__________ Date__________

Addess__________

Please ship at once_____cartons Edison Pickup Needles, containing 100 packages of 50 needles each. List price $10.00, on which your usual trade discount is to apply.

Dealer__________

Address__________

Electric Pick-up Needles
These Needles Are Best Suited for the NEW EDISON RADIO PHONOGRAPH COMBINATION
Loud and Extra Loud Needles are not recommended
TRADE MARK Thomas A. Edison
SAMPLE

3-162. In 1928, Thomas A. Edison, Incorporated found that its new radio-phonographs were subject to distortion when loud or extra-loud needles were used. Accordingly, the company supplied a lighter needle for use with these machines. Shown is a sample package of these needles, with a letter promoting them to dealers. *Courtesy of the Charles Hummel collections.* (Value code: VR with letter)

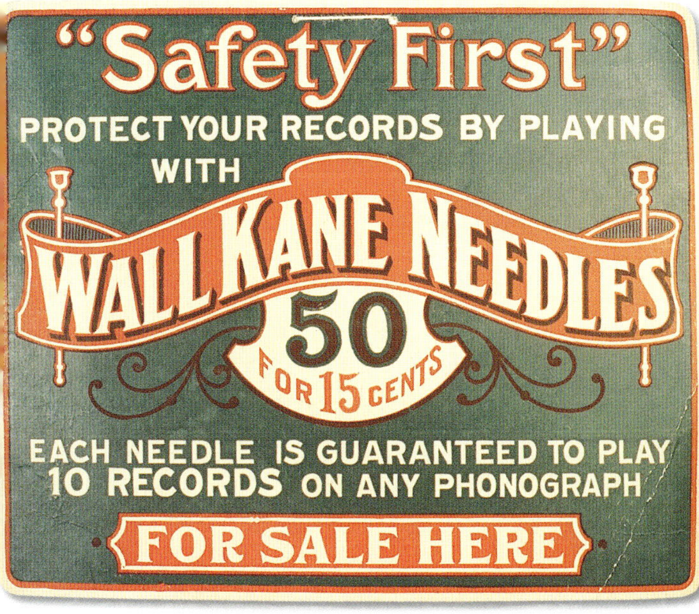

3-163. This 10 1/2" x 8 5/8" dealer hanger advertised Wall-Kane needles. It is ironic that a company which admonished "Safety First" (protecting ones records) would advise playing 10 records with the same needle! *Courtesy of Norm and Janyne Smith.* (Value code: D)

3-164. A companion piece to the hanger in the previous illustration was this Wall-Kane counter display with paper needle boxes. Illustrated in the August 1919 issue of the *Talking Machine World*, this display included 60 packages and was offered to dealers for $4.60 ($4.90 in far west). Retail was $9.00. Sold by the Progressive Phonographic Supply Company, Inc., New York City. 10 1/2" x 9 3/4". *Courtesy of Walter and Carol Myers*. (Value code: E)

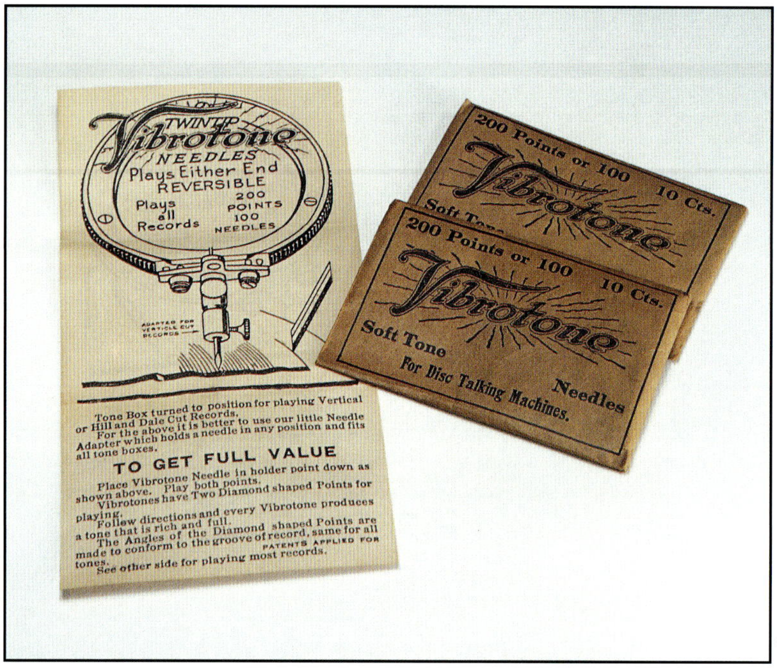

3-165. If double-faced disc records were a good idea, why not double-ended needles? The "Vibrotone" needles ("Twintip") seemed to make sense, until the ad copy mentioned vertical cut records. Edison and Pathé discs (designed to play with specifically shaped jewel styli) would not fare well beneath the "Diamond shaped Points…made to conform to the groove of record, same for all tones." One size, unfortunately, did not fit all. *Courtesy of Walter and Carol Myers.* (Value code: H)

3-166. If it's not an eggbeater, it must be the "Meltrope," a British gadget meant to sharpen steel needles. The sharpening kit included spare abrasive discs and collets. *Courtesy of Walter and Carol Myers.* (Value code: F)

3-167. Dust, doll-up, and dispense. The 3 1/2" record cleaner to the left carries seasonal greetings. The 2" Victor advertising mirror features an unusual border around the record. To the right is an HMV ("His Master's Voice") metal needle dispenser measuring 2 3/4" in diameter. *Courtesy of Walter and Carol Myers.* (Value code: E)

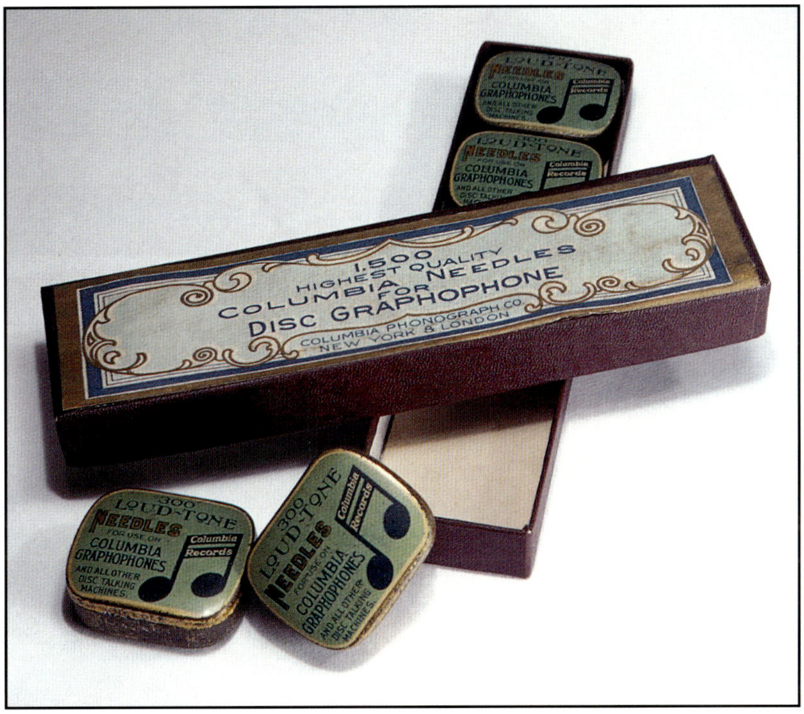

3-168. This rather elegant package contained five tins of Columbia Graphophone needles, circa 1910. *Courtesy of Walter and Carol Myers.* (Value code: G each)

3-169. Three needle tins, two of which (the "Dance Tone" and the "Landay") are also record cleaners. The large orange Columbia tin contains an assortment of 600 needles! *Courtesy of Walter and Carol Myers.* (Value code: F)

3-170. This enameled HMV needle sharpener was another device for sharpening steel needles. Such devices did not enjoy much popularity in the United States, whether by national temperament or better median income. *Courtesy of Walter and Carol Myers*. (Value code: E)

3-171. This British Columbia fiber needle cutter may give the fastidious collector a chill, since the instructions recommend screwing it to the machine. *Courtesy of Walter and Carol Myers*. (Value code: F)

3-172. Colorful advertising stamps for the German Beka record company. *Courtesy of Wilfried Sator*. (Value code: G each)

3-173. A group of stamps advertising a variety of machines, including the earliest metal-grilled version of the Victrola "IV." *Courtesy of Wilfried Sator*. (Value code: G each)

3-174. An unusual hexagonal box containing six triangular tins of "Violin" needles. *Courtesy of Wilfried Sator.* (Value code: VR)

3-175. An HMV wooden needle box with the trademark featured prominently on the lid. *Courtesy of Wilfried Sator*. (Value code: E)

3-176. A very clever needle dispenser. By sliding either one of the vertical magazines up and down, a new needle is left perched on the ledge at the top (look closely). The two cups are for used needles. *Courtesy of Wilfried Sator*. (Value code: VR)

3-177. A 12" x 10" store hange advertising German "Marschall" needles. *Courtesy* *Wilfried Sator.* (Value code: E)

3-178. This unusual Victor fiber needle cutter allowed the user to choose from six different cutting blades to produce varying "tones" – a concept bridging on absurdity. *Courtesy of Bernie and Katie VonThaer*. (Value code: E)

3-179. The reverse of the six-bladed Victor needle cutter. *Courtesy of Bernie and Katie VonThaer.*

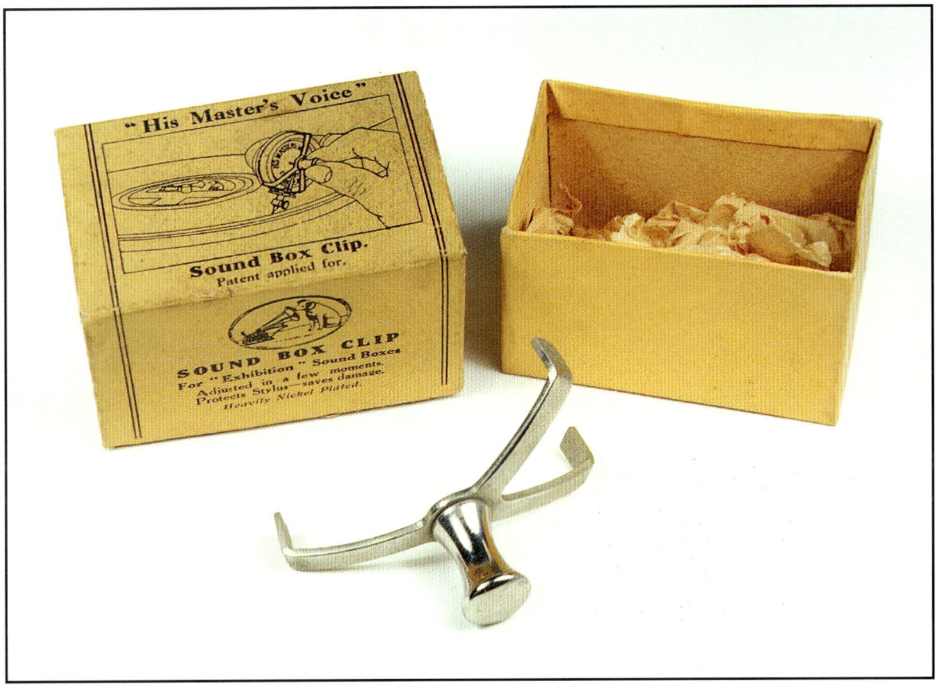

3-180. This British HMV "Sound Box Clip" afforded the user a better element of control when engaging the record. Not surprisingly, the container notes that this device "Protects Stylus – saves damage." Presumably, the record was the true beneficiary of this gadget. (Value code: G)

3-181. A slightly different (and less graceful) approach was taken by this gadget, whose only marking is "Victor Large Hole." *Courtesy of the Scott and Denise Corbett collection*. (Value code: F)

3-182. This rare Edison disc is a limited pressing of a speech given by Thomas Chalmers to Thomas Edison in 1927, commemorating the 50th anniversary of the invention of the Phonograph. Chalmers hails Edison for his glorious achievement with justifiable pride – he was an Edison employee! *Courtesy of Tom and Virginia Hawthorne.* (Value code: VR)

3-183. Never let it be said that the resourceful Brits lacked the wits for creating really arcane contraptions. The World "Record Controller," made by World Record, Limited of London, was a device that could be fitted to practically any disc talking machine, circa 1920. The World catalogue explained, "The outstanding feature of this invention is that it enables a gramophone record of equal size [to a conventional 12" diameter disc], to give three to five times the duration of entertainment." World intended to capitalize on a well-known inconsistency in disc record design, that continued to be problematic all through the 78 and later the LP eras. Unlike a cylinder, the *surface speed* of a disc record changes as the soundbox tracks toward the center. Although the *rotational* speed remains constant, the distance which the needle must travel in one revolution near the outside circumference of the record is considerably greater than it is nearer the label. Furthermore, the sound quality tends to diminish at the inner grooves where the *surface speed* is slowest. The "Record Controller" acted as an external speed regulator (it had a miniature governor built into it). It kept the *surface speed* of the record constant, effectively slowing down the normal *rotational* speed at the start of the grooves, speeding it up as the record played. This, of course, fulfilled the company's claim that their records (specially recorded to the specifications of the device) played longer. We see here the "Record Controller's" constituent parts, including a white celluloid "selector," which insured correct alignment of the device's large "tracking" wheel with the soundbox needle. (Value code: E)

3-184. Long before the publicizing of Carpal-Tunnel Syndrome, some users of the compact Mikiphone found its stiff, uncomfortable winding key to be unsatisfactory. This collapsible crank was just the ticket. *Courtesy of Garry James*. (Value code: C)

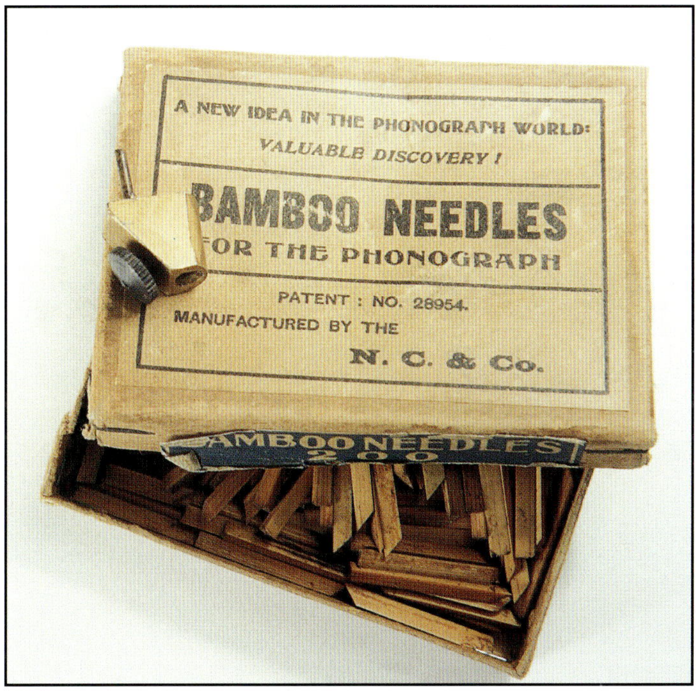

3-185. Around 1909, the use of bamboo and "fiber" needles became popular. Offering a softer tone, less scratch, and less record wear than steel needles, sales of fiber needles escalated. An additional advantage was the ability to sharpen these points with specially designed needle cutters. In the United States, the B&H Company successfully marketed this product. Victor entered the fiber needle business through the simple expediency of buying the B&H Company in 1910. This 2"x 3 3/4" box contains 200 bamboo needles and a chuck adapter to convert older soundboxes to accept the triangular-shaped bamboo or fiber needles. The patent number suggests that this package was not meant for American sale. *Courtesy of George Glastris*. (Value code: G)

-186. This 7" wood-backed record cleaning brush spoused a sentiment that many of today's collectors vould support: "A Victrola Will Brush Away All Your roubles." *Courtesy of George Glastris*. (Value code: G)

3-187. "Saves Records, Saves Needles, Eliminates Friction." This 2 5/8" diameter tin of disc record lubricant from England seemed to offer unlimited life to recordings. In practice, this oil probably harbored dirt and dust, promoting wear and shortening record life. *Courtesy of the Scott and Denise Corbett collection.* (Value code: F)

3-188. The end of World War I brought a wave of patriotic fervor that even found its way into the aesthetics of disc records. These two 10" Vocalion discs were meant to catch the eye as well as the ear. The example on the right contains a recording of fireworks explosions and noisemakers, as well as a celebratory label. *Courtesy of Michael F. Khanchalian*. (Value code: VR)

3-189. The Victor company must have greatly approved of the impression created by this charming sheet music: that the Victrola was a more robust instrument than the external horn machines that preceded and, for a time, co-existed with it. *Courtesy of Alan H. Mueller.* (Value code: F)

3-190. The cabinetry of phonographs became so enmeshed in popular culture that non-related items were sometimes made to resemble talking machines. Shown is a 6" long jewelry box marked "Ivoroid" (actually, celluloid) which resembles a console phonograph of the late teens or early 1920s. *Courtesy of the Jerry Koch collection*. (Value code: F)

3-191. Three effigies of phonographs. The example to the left is a stylized celluloid creation with a sliding door. The center and right examples are very close copies of Victor Victrolas from the first half of the 1920s. The largest measures 6" high. *Courtesy of the Tom and Sandi McCarthy collection.* (Value code: F)

3-192. A close up of another example of the larger Victrola seen in the previous illustration. The spit and image of a Victrola "230," this cast metal piece is marked "Rehberger Mfg. Co., Chicago." Interestingly, the lid carries a miniature replica of the Victrola decal found beneath millions of full-size Victrola lids. *Courtesy of Don E. Menez.*

3-193. As the world entered the Jazz Age, record companies vied with each other to supply the latest music to its customers. Notably lagging in this regard was the Edison enterprise, which took marching orders from its septuagenarian self-styled music critic pictured on this record sleeve. The faithfulness of the Edison disc recordings is legendary, but the claim above the great inventor's likeness was, unfortunately, only wishful thinking. *Courtesy of George Glastris*. (Value code: H)

3-194. In 1926, the Edison company was on the horns of a dilemma: use its considerable pool of talent and resources to innovate, or remain crushed under the weight of its immense reputation, obsolete product, and outmoded business strategies. Like all plans destined for failure, Thomas A. Edison, Incorporated embarked on a halfway measure. The firm developed a "Long Play" version of the "Diamond Disc" Phonograph (40 total minutes of music from a 12" disc) that required special gearing, special records, and a special reproducer. A great idea – followed by the introduction of four identically drab "Long Play" machines, and a short list of perfectly deadly records. End of story. Not surprisingly, this LP reproducer remained on the shelf at the Levis Music Store in Rochester, New York until the 1960s. *Courtesy of Alan H. Mueller*. (Value code: D)

3-195. This die-cut card depiction of an upright talking machine offered the prospective customer an unusual amenity: a room thermometer. A grommet was once affixed to the hole at the top, by which the card could be mounted on a wall. (Value code: G)

3-196. Few areas of study offer the variety and ingenuity represented by the accessories and contraptions developed for disc talking machines. The juxtaposition of a shop selling Gramophones (and clocks) in a centuries-old half-timbered building is evidence of the continuous renewal of human enterprise. (Value code: H)

Four:
Cabinets and Storage

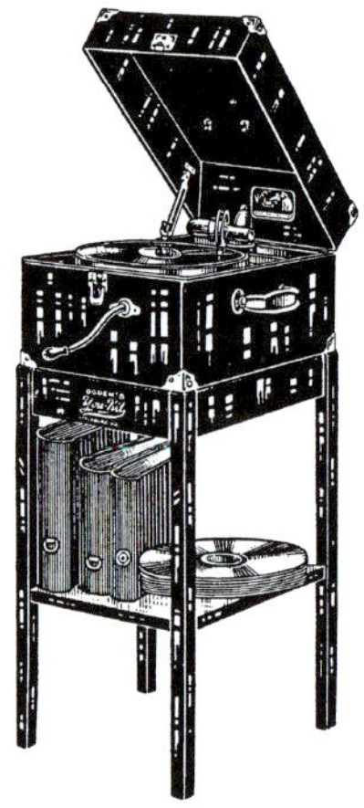

Much has been written about the advantages the disc record posed to prospective buyers versus the cylinder. It will be evident to the readers of this chapter that the same amount of cabinet space would hold considerably more discs than cylinders. The fragility of "wax" cylinders, and the relative durability of shellac 78s also made discs more attractive to the public. Yet, even the widespread availability of virtually unbreakable celluloid cylinders during the period 1907-1912, could not level the playing field.

One aspect of competition between the two formats that has not been explored, however, is the inability of the ingenious minds which created the panoply of cunning contrivances already described in this book to invent a *selective* system for the management of domestic accumulations of cylinder records.

Disc owners were offered a number of handy mechanisms by which a record could be located and easily extracted from a storage cabinet. In 1907, the Automatic Disc Holder Company of New York City advertised an apparatus as follows, "You have but to *press the key* and the mechanism rolls it [the record] slowly out into your hand." Edward F. Pooley patented his Pooley System (No.980,124, December 27, 1910), and manufactured, "The Cabinet That Finds and Hands the Record to You." Discs were selected from the vertical storage compartment with the help of an index, and dispensed by a lever into the hand (see illustration 4-106).

The Wood Manufacturing Company of Los Angeles ("Simply Push the Button") produced cabinets fitted with a push-button retrieval apparatus, and many of the company's systems were installed in the record storage cupboards of Columbia Grafonolas (see illustration 4-120). The "Eject-o-file" advertised by the Peabody Piano Company of Baltimore, Maryland, in 1918 was designed to fit the cabinets of specific talking machines (Model "150-E" was for an Edison Diamond Disc "C-150"). A simple rocker tilted each record out of its indexed slot. A similar product, the "Crip-N Ejex File" made by the Crippen Record File Corporation of Rochester, New York, was advertised as follows in the December 1919 *Talking Machine World*, "Every Record Instantly Available… Any desired selection is always at the finger's tip… With a 'Crip-N Ejex,' which is easily installed in all makes of phonographs, real service, economy and efficiency is provided."

Nowhere was there a convenient mechanism for selecting and dispensing cylinders. The task was simply too complex to make it practical. Only in the realm of coin-operated phonographs could an indi-

vidual cylinder be indexed, isolated, and played by mechanically-assisted means. The complicated works required to accomplish this end were beyond what most homeowners could afford or keep in adjustment.

Beginning around 1907, coin-in-the-slot instruments such as the Regina "Hexaphone" (six selections), the Multiphone (up to 24 selections) and some obscure, never-seen oddities such as the Multinola (16 selections, briefly advertised in 1911 by the American Multinola Company, Cleveland, Ohio) were contrived or manufactured.

The only domestic cylinder instrument that provided any semblance of selectivity was the Autophone of the early teens, a machine in a large floor-standing cabinet which was about as big as any private residence could accommodate, and held only 12 records on a carousel. Clearly, the cylinder had insurmountable physical disadvantages. Although some present-day collectors may find cylinder storage cabinets the most interesting or unusual, the original owners of these artifacts were never afforded the same level of convenience as their neighbors who owned disc talking machines.

4-1. The elegantly simple design of this cylinder record carrying case would make it an acceptable addition to the parlor. To find so basic an object elevated to an artistic level is unusual. Double-sided, 25 flannel-covered pegs to the side, 17 1/2" wide, 11" deep and 20 1/2" high. *Courtesy of the Charles Hummel collections.* (Value code: VR)

4-2. The sophisticated tambour-fronted storage spaces dispensed with inefficient hinged doors, but allowed access to only one side of the carrying case at a time. *Courtesy of the Charles Hummel collections.*

4-3. Between 1889 and January of 1897, the Columbia Phonograph Company was headquartered in Washington, DC. The simple box shown here contains brown wax cylinders from this period. Note the paper ribbon encircling the base of the cylinder in the foreground: a "title slip" that couldn't be mixed up or easily lost. (Value code: VR)

4-4. This cylinder record storage/carrying case may be seen in an illustration of the Bettini laboratory pictured in Roland Gellatt's *The Fabulous Phonograph*. The manufacturer is unknown. The thought of trusting the brass catch with the safety of brown wax cylinders may give modern collectors the chills. Evidently, the original owner had some misgivings of his own, as each wooden peg has been carefully wrapped with cotton and tied with thread. (Value code: VR)

4-5. This canvas-covered carrying case from the late 1890s holds 36 brown wax cylinders. The variation in the color of the records, from light tan to dark chocolate brown, was quite common at a time when individual batches of metallic soap or "wax" could differ considerably in content. *Courtesy of Domenic DiBernardo.* (Value code: E, case only)

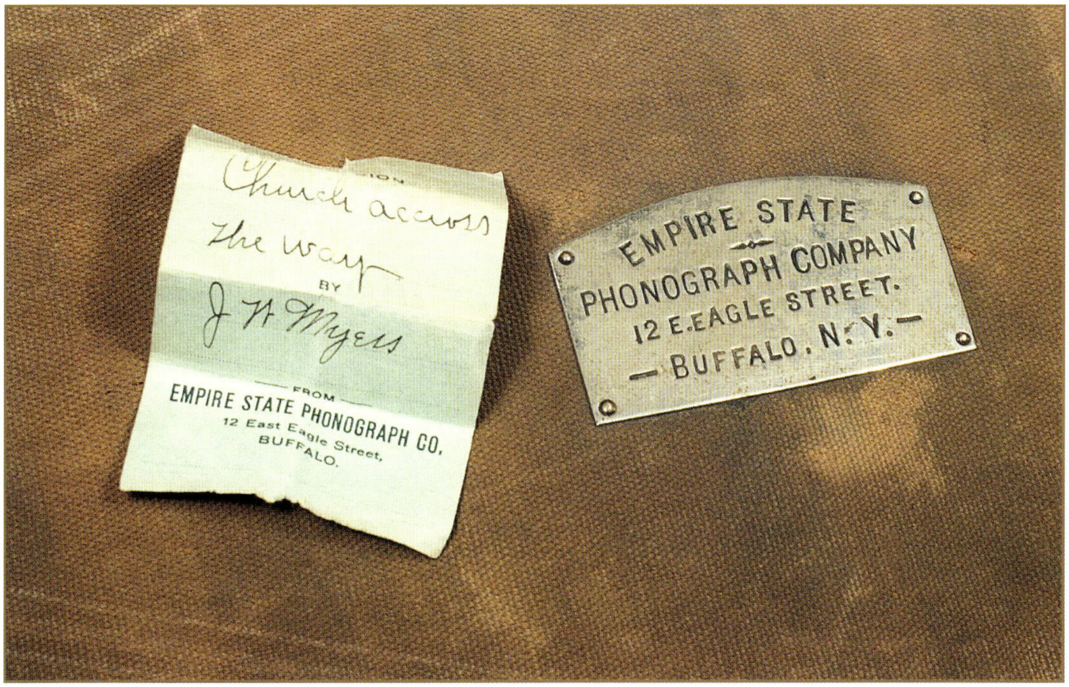

4-6. Atop the carrying case seen in illustration 4-5 is this nickel-plated dealer plate. Shown also is one of the cylinder title slips that accompanied each of the records. *Courtesy of the Domenic DiBernardo collection.*

4-7. Two "Lynn" cylinder record carrying boxes showing two typical styles of the late 1890s – pasteboard and canvas covered wood. The "Lynn" record index displayed on the right is just the kind of item that usually gets lost. *Courtesy of Michael Sims.* (Value code: G, F, boxes only)

4-8. An Edison "Spring-Motor" Phonograph is ensconced in a mahogany cabinet containing three drawers (12 pegs per drawer) and "glazed lid." This style of cabinet strongly resembles early Hawthorne & Sheble models, but with an uncommon delicacy. Measures 30 1/4" wide x 17 3/4" deep x 38" high. *Courtesy of the Sanfilippo collection*. (Value code: VR)

4-9. Record cabinets designed to accommodate the Edison Class "M" (later, "Balmoral") mechanism are unusual. This example features five drawers for cylinders (30 per drawer), plus a small drawer for accessories. The "glazed lid" hinges backward for access to the mechanism. Measures 24" wide, 17 3/4" deep, and 43 1/4" high. *Courtesy of the Sanfilippo collection*. (Value code: VR)

4-10. During the first decade of the twentieth century, the cylinder record storage cabinet came within reach of vast numbers of talking machine owners. However, during the 1890s, cylinder cabinets were less common, particularly a mahogany showboat such as this one. Elegantly finished on all sides, this piece can hold 217 records on wooden mounting pegs. It's quite possible that it once supported a Class "M" Edison electric Phonograph, an equally imposing instrument. 30 1/4" wide, 27" deep and 33 1/2" high. *Courtesy of Nina and Dave Heitz*. (Value code: VR)

4-11. The "Grand" cabinet was manufactured by Hawthorne & Sheble of Philadelphia, and first catalogued by Columbia in1898 at $50.00. After 1901 (when it was offered for $35.00), it disappeared from American catalogues (including Hawthorne & Sheble's own), but can be found as late as 1906 in Columbia's British offerings. Consistently described as being "made of black oak" (or "dark oak" in the British catalogue), this example is, surprisingly, executed in mahogany. The drawers could accommodate 174 standard size cylinders, or 42 "Grand" cylinders. *Courtesy of the Domenic DiBernardo collection*. (Value code: VR in mahogany)

4-12. This Columbia Graphophone Type "B," circa 1900, has been housed in a professionally built "hand cabinet" in which the horn is cleverly secured within the interior of the lid. The problem of how to keep the horn from getting misplaced was thereby solved. *Courtesy of John D. Robles.* (Value code: B)

4-13. This custom designed deal box was discovered with a Berliner "Improved Gramophone," circa 1898. On the left is a space for the soundbox and other small parts. The record storage on the right is of particular interest. We may observe first-hand the impact and inevitability of the disc record. A dozen Berliner discs could be stacked in an area suitable to hold only a handful of cylinders. The scrap of newspaper indicates when the machine was put away in the attic. *Courtesy of John P. Andolina Jr., the Early Sound Man.* (Value code: VR)

4-14. This neatly dovetailed oak storage box for 7" discs stores them horizontally, which is unusual. This arrangement, coupled with careless handling or material failure could be a nightmare for early disc collectors. Fortunately, the steel handle is far less likely to break than the leather models commonly seen on other disc storage boxes of the time. Measures 8 3/4" wide x 7 3/4" deep x 9 1/8" high. *Courtesy of the Sanfilippo collection.* (Value code: VR)

4-15. Both the Berliner Gramophone Company and the Universal Talking Machine Company ("Zonophone") endorsed or distributed wooden and pasteboard record cases for 7" discs. This unusual canvas covered model has not been found in contemporary catalogues. *Courtesy of Michael F. Khanchalian.* (Value code: C)

4-16. In its opened position, the "Zonophone Record Case" displays an index with typical American selections inscribed by the original owner. *Courtesy of Michael F. Khanchalian.*

4-17. A Hawthorne & Sheble 5" diameter cylinder double carrying case. In 1901 this item retailed for $12.50, when each "Concert" or "Grand" cylinder cost $1.00. (Value code: E)

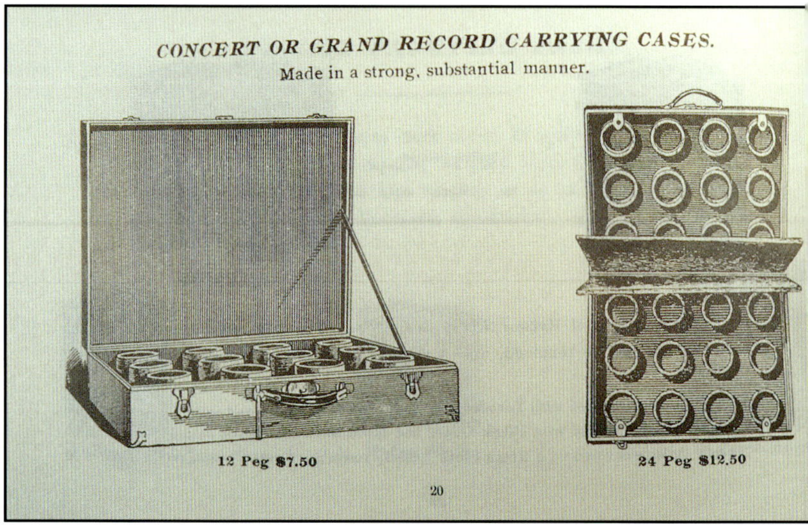

4-18. The carrying case shown in the previous illustration, as featured in the H&S catalogue.

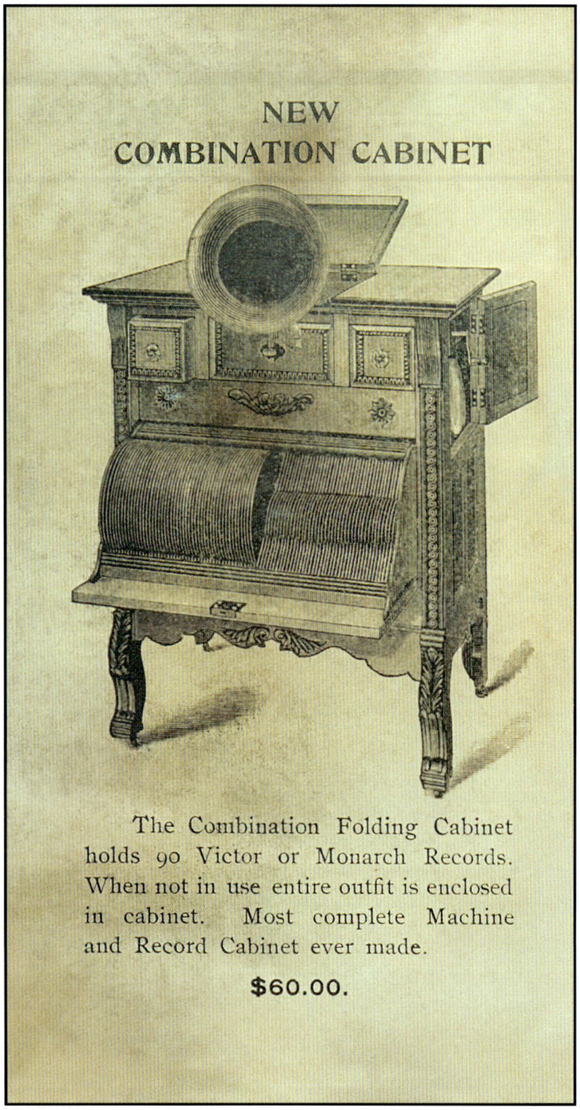

4-19. This 3 1/4"x 6 1/4" page from an Eldridge Johnson catalogue of 1901 shows a stunning (and sadly undiscovered) combination "Machine and Record Cabinet." Expensive for the time at $60.00, very few could have been sold.

4-20. The 1901 Hawthorne & Sheble catalogue showed a "Folding Oak Stand" which looked much like a sewing table. The advantages of this stand would have appealed primarily to exhibitors, few of whom were using Gramophones or Zonophones in 1901, because of the difficulty of appending large amplifying horns to them.

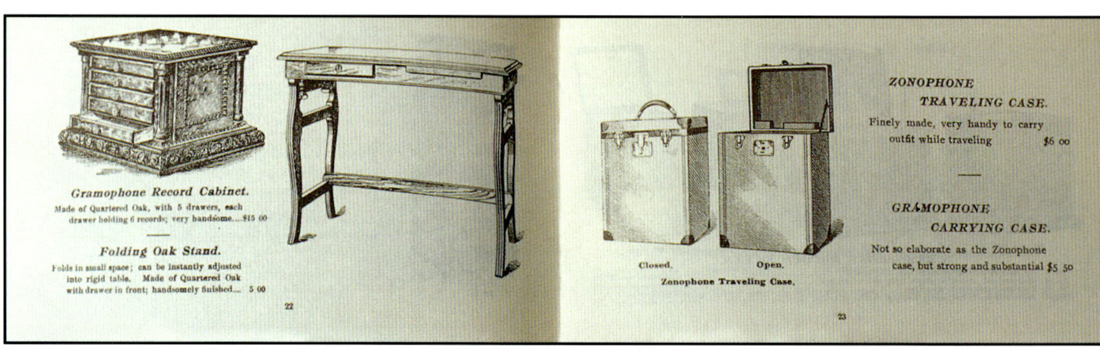

4-21. The same "Folding Oak Stand" in reality. The additional braces imparted stability, which is noticeably lacking without them. Similar stands occasionally turn up without hinged legs, and some have patent dates from the teens impressed in the wood, suggesting a long period of manufacture with various applications. This hinged-leg example supports a Zonophone Model "C" from 1900. (Value code: C)

4-22. A Hawthorne & Sheble No."100" Disc Machine and Record Cabinet as featured in its 1903 catalogue. With a capacity of 100 records, this cabinet listed for $14.00. Measures 21 1/2" wide x 17 3/4" deep x 36 3/4" high. *Courtesy of the Sanfilippo collection*. (Value code: A)

4-23. The conveniently pivoting needle receptacle features an enameled brass plate bearing the Hawthorne & Sheble attribution. *Courtesy of the Sanfilippo collection.*

4-24. Gianni Bettini has an exalted reputation for having produced cylinder recordings of exceptional quality and content. Today's collector views Bettini cylinders with the reverence of sacred vessels. Shown here is a pasteboard box bearing Bettini's distinctive label, circa 1902. (Value code: VR)

4-25. This modest wooden carrying case for cylinder records, circa 1903, is made noteworthy by the vibrant paper label on its upper surface. Similar artwork was used in the United States, and the French text is a literal translation: "Columbia cylinders are best. On sale here." *Courtesy of Philippe Le Ray.* (Value code: D)

4-26. A Columbia pasteboard record holder fitted with very interesting pegs. The body of each peg is made of tinned sheet metal, capped by wood. Wax cylinders could have benefited from this type of storage, had it been more widely used. Metal does not attract and hold moisture, or promote the growth of mold on the records. *Courtesy of Steve and Ellie Saccente.* (Value code: F)

4-27. This oak cylinder record carrying case with an elegant brass bail pull also employed metal pegs. Despite their advantages, they were undoubtedly more expensive to manufacture than cardboard, and are consequently rare today. *Courtesy of the Domenic DiBernardo collection.* (Value code: E)

4-28. A number of configurations were employed to store and transport cylinder records efficiently. One of the least complicated is this two-tiered carrying case housing 36 records, from the Columbia Phonograph Company (branch) of Paris. It's interesting to note that "Paris" appears before "New York" in the logo, a predictable reverse of the wording used on objects for United States consumption. *Courtesy of Yves Rouchaleau.* (Value code: D)

4-29. Looking more like a shoeshine kit than a cylinder record case, this three-tiered example was probably the creation of a talented carpenter. Each box contains 18 wooden pegs and, once stacked and hooked together, the entire case was held rigid by brass lugs. The interior bottoms of each level are lined with felt. *Courtesy of Richard G. Allen.* (Value code: D)

4-30. This oak three-tiered disc cabinet suggests a date prior to 1903, since it accommodates only 10" discs on its individually grooved shelves. Measures 14 1/2" square x 38 1/4" high. *Courtesy of the Sanfilippo collection*. (Value code: B)

4-31. The Englewood Company of Chicago distributed disc talking machines and records manufactured under its name by firms such as Columbia (the "Englewood Musicalphone") and the International Record Company ("Nightingale Parlor Grand Records"). Shown here are two cardboard containers that accompanied a "Musicalphone." One to hold records, and the other to house the soundbox. *Courtesy of Joan and Robin Rolfs.* (Value code: VR)

4-32. The disassembled parts of a Victor "Monarch" complete with its "Taper Arm" and horn were stored and transported in these sturdy felt-lined containers. A variety of companies offered these accessories, notably Hawthorne & Sheble of Philadelphia. *Courtesy of Tom Sideleau*. (Value code: C, case only)

4-33. The sober exterior of this cylinder record cabinet belies the delicate and imaginative design of the internal fittings. The combination of wood turnings and rattan is made more remarkable when one considers that all of it will be hidden from view when the 165 cylinder records are put in place. Measures 20 1/2" wide, 16 1/2" deep and 34 1/2" high. *Courtesy of the Gfell family collection.* (Value code: B)

4-34. This "Perfection Pivot Cylinder Record Cabinet" was sold by the Douglas Phonograph Company of New York City. An advertisement in the August 1905 issue of *Talking Machine World* stated: "The most practical cabinet constructed for this purpose. Drawer shelves revolve… Hollow turned wooden pegs firmly screwed to drawer." The ad showed a cabinet with similar shelves, though flattened on the front edge, which limited the capacity to 28 pegs per drawer. Douglas later improved the shape of the shelves (as seen here), creating a more efficient use of space (38 pegs per drawer). Measures 26 1/2" wide, 15" deep and 40" high. *Courtesy of Charles McCarn.* (Value code: A)

4-35. Unlike disc cabinets, few cylinder record storage cabinets were equipped with a slide-out shelf on which to place records. Here, a shelf of relative substance provides a convenient platform for sorting cylinders. 24" wide, 20" deep and 29" high. *Courtesy of Doug Defeis.* (Value code: A)

4-36. A mahogany cylinder cabinet of fanciful design, featuring flaming torches at the four corners. Inside are five sliding trays that hold a total of 218 cylinder records. 23" wide, 19 1/2" deep and 36" high. *Courtesy of the Sanfilippo collection.* (Value code: A)

4-37. Victor H. Rapke, a talking machine jobber, developed a system for storing commercial inventories of Edison cylinder records in trays. His idea was first mentioned in the *Edison Phonograph Monthly* in December 1903. The trays initially held six records stacked one on top of the other with lids facing out. Rapke later instituted a horizontal plan, like the one shown. The trays could be placed on shelves or any surface within a commercial establishment. The most important feature was the large label that identified the selection being stocked. The April 1905 issue of the *Edison Phonograph Monthly* noted that Rapke sold a complete set of gummed labels (covering 2200 titles in the Edison two-minute record catalogue) for $1.50. Mention continued to be made in the *EPM* of this practical storage system until February 1916. *Courtesy of the Scott and Denise Corbett collection.* (Value code: H)

4-38. Evidently the work of a skilled local carpenter, this cylinder record rack might have been attached to the wall of a shop handling phonographs during the first decade of the twentieth century. The capacity of 105 records could represent a respectable variety of titles, and the lid-out design made locating them easy. Measures 46 1/2" wide, 4" deep and 31" high. *Courtesy of Domenic DiBernardo.* (Value code: A)

4-39. A cabinet specially designed to support a Victor "VI" talking machine. Note the pillars, apron, and cabriole legs. Interestingly, the doors open from each end, a design reminiscent of some cabinets from the 1890s. Measures 29" wide x 22" deep x 34 1/2" high. *Courtesy of the Sanfilippo collection.* (Value code: VR)

4-41. The Victor "VI" was the company's most luxurious model, but some well-heeled individual wanted more – much more. The Douglas Phonograph Company of New York City installed Victor components in custom cabinets that made the stock gilt hardware and polished mahogany of the over-the-counter "VI" look unexceptional by comparison. 20 1/2" wide, 16" deep and 47 1/2" high. (Value code: VR)

4-40. A cabinet similar to that in the previous illustration shares pillars, apron, and cabriole leg designs, but with double doors which open simultaneously by means of an arrangement of rods below. It is possible that these two cabinets were manufactured by the same firm. Measures 21 1/2" wide x 16 1/2" deep x 36" high. (Value code: VR)

4-42. The gilded fittings of the Victor "VI" dim before the luminously embellished cabinet.

4-43. Eighteenth century pastoral scenes – just right for the drawing room. This Douglas cabinet and the one shown in illustration 4-34 are recognizably similar in overall construction.

4-44. In the Douglas cabinet, records are stored on sliding shelves. The July 1907 *Talking Machine World* announced, "A New Disc Cabinet… has Two Sliding Record Racks that pull out so that records are easily accessible… total capacity, 150 12-inch records."

4-45. A very imposing disc record cabinet in the Empire style. The bow front, carefully arranged pattern of the mahogany veneer, and impressive capitals and pediments all bespeak elegance. 23 1/4" wide, 16 7/8" deep and 38 1/2" high. *Courtesy of Charles McCarn.* (Value code: A)

4-46. The doors of the Empire-style cabinet open in tandem, and the interior is arranged with conventional dividers. Note the stamped brass needle cups at the top of the right hand door. This cabinet was designed to support a fancy external horn talking machine, such as the Columbia "BD." *Courtesy of Charles McCarn.*

4-47. The sinuous design of this oak record cabinet suggests the lines of a "serpentine" Reginaphone. The extendable slide at the top provided a handy platform for resting discs or making entries in a record index. 23" wide, 18" deep and 34" high. *Courtesy of the Sanfilippo collection.* (Value code: A)

4-48. The interior of an identical cabinet executed in mahogany (perhaps best suited to the Columbia Disc Graphophone Type "BY") employs vertical wooden dividers. *Courtesy of Walt and Ellen Bellm.* (Value code: A)

4-49. A mahogany disc storage cabinet that combines an unusual mix of styles, from its incongruously delicate feet to the heavy plank that constitutes its top. The decorative columns, with their Ionic capitals and fluted shafts might complement a number of disc talking machines, such as the Victor "VI" or the Columbia "BNWM." New and used needle access is of the simplest type: two recesses gouged in the thickness of the door. 26" wide, 18 3/4" deep and 34" high. *Courtesy of Jen and Ken Flaherty.* (Value code: A)

4-50. This mahogany disc cabinet featured columns on the front corners and flattened ball-and-claw feet. Measures 27" wide x 18" deep x 35 1/2" high. *Courtesy of the Sanfilippo collection.* (Value code: A)

4-51. The design of this mahogany Queen Anne cabinet allowed convenient access to the cylinder records it stored. Each shelf of pegs within the interior framework of the doors pulls out. A very respectable number of records, 142, can be contained, despite space lost to the medium-height legs. The "back splash" or header at the top rear usually denoted a sheet music cabinet. Measures 22 1/2" wide, 16" deep and 42" high (to the top surface). *Courtesy of the Scott and Denise Corbett collection.* (Value code: A)

4-52. A sturdy-looking paneled oak cylinder record cabinet. The trays contain 240 numbered cardboard pegs on which records are stored. 24 1/2" wide, 20 1/2" deep and 32" high. *Courtesy of Steve and Ellie Saccente.* (Value code: B)

4-53. In October 1908, the Talking Machine Company of Chicago announced "The Cabinet That Matches." That month's issue of the *Talking Machine World* carried an advertisement that stated, "A cabinet for each Victor machine, and each machine absolutely fits... We make these cabinets, you will recognize our guarantee." The "VI" cabinet sold for $40.00, and listed a capacity of 220 records. Measures 22" square x 35 3/4" high. *Courtesy of the Sanfilippo collection.* (Value code: VR)

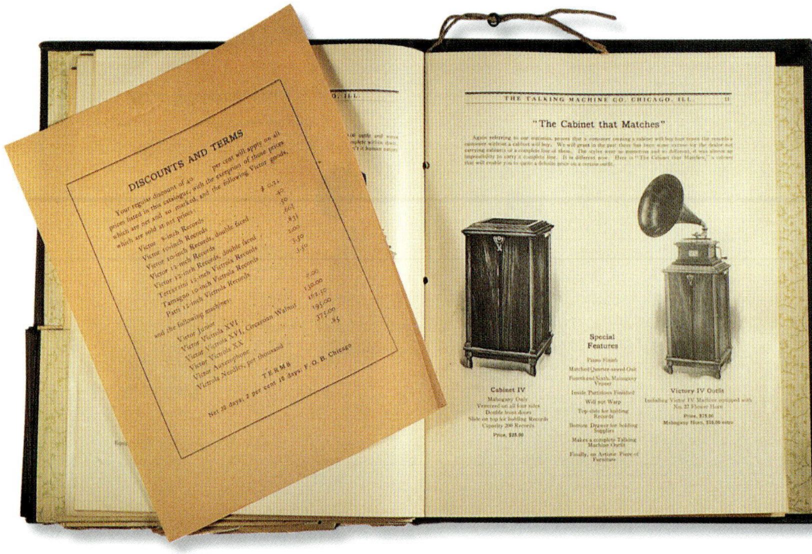

4-54. An important feature of "The Cabinet That Matches" was the so-called "continuation" top. Each cabinet was specially designed to resemble in appearance a specific model of disc talking machine. The upper molding of "The Cabinet That Matches" fit snugly around the instrument, giving the impression of a single piece of furniture. Here, from around 1907, is a catalogue and wholesale price list.

4-55. This highly-decorative cabinet is the type sold in Great Britain and Europe during the first decade of the twentieth century to fit posh external-horn Gramophones. A cabinet similar in style, but with ebonized finish, was used to support the "Melba" talking machine sold by the Gramophone Company, Limited. Measures 17" x 17" x 39" high. *Courtesy of Bruce and Linda Peterson*. (Value code: A)

4-56. An oak Herzog cylinder record cabinet of special charm. The bowed sides are imitated in the upper molding, which sags gracefully like ice cream melting down the sides of a cone. The doors are interconnected and open in tandem. Five shelves of pegs hold 36 records each. 24" wide, 20" deep and 34" high. (Value code: A)

4-57. A very tasteful cylinder record cabinet in oak with routed sides and a graceful skirt. Records are held on wooden pegs, 20 per drawer. 15" wide, 19 1/2" deep and 35 1/4" high. *Courtesy of Jen and Ken Flaherty.* (Value code: B)

4-59. The Herzog Art Furniture Company of Saginaw, Michigan first advertised this remarkable piece of furniture in the August 1907 *Talking Machine World* under the designation No."834" "De Luxe Concealed Horn Cabinet" ("holds Any Size Disc machine"). The September issue of the same publication announced that Herzog had abandoned the "De Luxe" nomenclature for legal reason in favor of "The Teddy." This name was also discontinued and by October 1908 the cabinet was known simply as the "Concealed Horn Cabinet." The concept was clever: a conventional external horn disc talking machine was secreted within a stately mahogany closet. *Courtesy of the Sanfilippo collection.* (Value code: VR)

4-58. Disc record indexes are not uncommon, but a cylinder index is an incredible rarity. Except for the interior lid surface of pasteboard storage boxes (which often had space for listing records), most cylinder phonograph owners used sheets of ordinary writing paper as an index. Illustrated, in mahogany, is the posh cabinet shown in Figure 4-56. Measures 4 1/2" x 10 1/2" folded.

4-61. A seldom-seen variation of the Herzog "Concealed Horn Cabinet" grille, made of mahogany-finished composition material. *Courtesy of the Sanfilippo collection.*

4-60. A Victor Model "0" (zero) fits cozily inside the "Concealed horn Cabinet," with space below for disc record storage. By closing the doors, the machine can be completely hidden during operation, except for a hint of amber horn behind the decorative composition grille. The April 1908 *Talking Machine World* featured an advertisement for a similar Herzog cabinet known as the "Cyl-o-phone," designed to accept a cylinder phonograph. The following month, *TMW* described the same unit as a "Cyclo-phone," before returning to the original terminology. Herzog belittled the competition in *frosty* language, "Do not buy Substitutes, Imitations, Refrigerators. Our Cabinets match any and all makes of Machines." 24 1/2" wide, 20 1/2" deep and 55 3/4" high. *Courtesy of the Sanfilippo collection.*

4-62. This interesting cabinet (circa 1908) from Lyon & Healy of Chicago adopted a lid design reminiscent of the first Victrola of 1906. Measures 18 1/2" wide x 17 1/2" deep x 37 1/2" high. See next illustration. *Courtesy of Sam Sheena*. (Value code: A)

4-63. Opening the lid reveals an Edison "Home" Phonograph. Each drawer holds 25 cylinder records. Note the partially opened sliding shelf above the drawers: most unusual in cylinder cabinets. *Courtesy of Sam Sheena*.

4-64. This Victor "V" is distinguished by the Allegrophone sitting atop it (see next illustration), as well as the "Cabinet That Matches" below. The "V" cabinet sold by the Talking Machine Company of Chicago retailed for $30.00. One might argue that the routing on the cabinet's columns does not precisely match the Victor's design. The same cannot be said, however, of the Allegrophone. The record cabinet measures 20 1/2" square x 37 1/4" high. *Courtesy of the Sanfilippo collection*. (Value code: VR)

4-65. The Allegrophone was advertised in the September, October, and November 1908 issues of the *Talking Machine World*. Manufactured by the National Allegrophone Company of Boston, Massachusetts, this device was offered in quartered oak for $12.50, or mahogany for $15.00. "Dignifies and Beautifies the Talking Machine. Is to the Talking Machine what the case is to the Piano." The company went on to point out the Allegrophone's advantages: "Conceals the running parts and eliminates all the noise except that necessarily transmitted through the horn. Can be adjusted in a minute, remains permanently and loses its identity in that of the machine. Has plate glass in the sliding top through which the operator may watch the progress of the needle. Protects delicate mechanism and records." For any Doubting Thomas, the company confided, "Today the Allegrophone is a new invention. Tomorrow you will see it on all Talking Machines. It will prove a boomer." Judging by the rarity of the Allegrophone today, we may assume that it proved something less than "a boomer." *Courtesy of the Sanfilippo collection.*

4-66. A mahogany finish disc cabinet, with three tiers suitable for 10" diameter records. Note the clever needle accessory on the door. The bottom of the used needle receptacle swings out, allowing it to be easily emptied. *Courtesy of Steve and Ellie Saccente.* (Value code: B)

4-67. The Rockford Cabinet Company of Rockford, Illinois produced this circular cabinet for cylinder records, shown here in quarter-sawn oak, measuring 23" diameter by 40" high. *Courtesy of the Babcock House Museum, John H. Perschbacher.* (Value code: VR)

4-68. An article about this Rockford cabinet in the October 1908 *Talking Machine World* reported, "The shelves are supported on individual pivots and any one may be swung out separately, thus permitting easy access to the records. The cabinet contains five shelves each capable of holding thirty-seven records, and is beautifully finished throughout." An illustration in the magazine depicted shelves with pegs, however the article continued, "For those who prefer to keep their records in their original cartons the shelves are [then] provided with supporting strips around the edges and each will then accommodate forty records [as shown here]." *Courtesy of the Babcock House Museum, John H. Perschbacher.*

4-69. At first glance, this oak cabinet appears as if it might contain storage for either cylinder or disc records. It measures 22" wide, 14 1/2" deep and 38 3/8" high. Note the unusual external door hinges. *Courtesy of Nina and Dave Heitz.* (Value code: A)

4-70. The cabinet interior contains shelves that pivot outward in a manner also employed by the Rockford company. Whereas Rockford's shelves swung out from the left, the shelves of this cabinet hinge from individual castings located on the right. *Courtesy of Nina and Dave Heitz.*

4-71. Louis Buehn & Brother of Philadelphia distributed cylinder and disc record cabinets. Shown here are several dealer sales catalogues circa 1907. As indicated in one of the illustrations to the right, piano rolls could be stacked on a shelf meant for disc record albums. *Courtesy of Tom and Joanna Broome.*

4-72. The Musical Echo Company, Victor dealers of Philadelphia, also specialized in both cylinder and disc record storage items. These flyers date from 1909. *Courtesy of Tom and Joanna Broome.*

4-73. This is the type of record album advertised as "for sale by all Victor dealers" by the Musical Echo Company. Disc records at this time were single-faced. In this particular album, two records fit back-to-back in one sleeve, with circular cutouts to make the labels visible. L. J. Gerson, manager of Musical Echo, apparently applied for a patent regarding album storage. He did receive a patent, but for a spring counter-tension device to relieve the weight of the Victor soundbox on the record (No.962,565). The most surprising personality associated with the company, however, was F.W. Woolworth, the dime-store magnate. As reported in the January 1907 issue of *Talking Machine World*, Mr. Woolworth held a controlling interest in the Musical Echo Company. *Courtesy of the Scott and Denise Corbett collection.* (Value code: G)

4-74. The same two-disc-per-sleeve design is seen in this vividly decorated album for single-faced 7" diameter German Gramophone records. *Courtesy of Michael F. Khanchalian.* (Value code: F)

4-76. Distinguishing this Edison record shipping crate, with its corrugated cardboard inserts, is the lively insignia of the jobber, W.H. Thorne of New Brunswick, Canada. Thorne's name first appeared in the December 1906 issue of the *Edison Phonograph Monthly*, and remained among the gradually dwindling list of jobbers until at least January 1917. *Courtesy of Tom and Sandi McCarthy.* (Value code: E, box only)

4-75. During the first decade of the twentieth century, wire store fixtures were manufactured to hold both cylinder and disc records. These lightweight racks were designed to be practical, not decorative – but this empty unit betrays a certain geometric appeal. When filled, the 200 cylinder records appear to float in midair. 27" wide x 52 1/2" high. *Courtesy of Michael F. Khanchalian.* (Value code: B)

4-77. An oak disc record cabinet with "continuation top." The shaped, recessed side panel is highly unusual. The pillars flanking the front, with their fluted columns and Ionic capitals, would blend easily with a number of external horn disc talking machines from the 1905-1910 period. The general appearance suggests the Talking Machine Company of Chicago's "Cabinet That Matches" line. 19 3/4" wide, 19 3/4" deep and 35 1/2" high. (Value code: A)

4-78. This Columbia "BII" Disc Graphophone (with a "No. 2" wooden horn, refer to illustration 1-95) rests on an unmarked "continuation top" record cabinet. The rather plain sides are relieved by a stylish paneled door and small apron. Measures 18 1/2" square x 34" high. *Courtesy of Freddie and Jeanette Blair.* (Value code: A, cabinet only)

4-79. The patent for this unusual cylinder record cabinet with retractable bent-wood horn support was awarded to Samuel Solomon on September 15, 1908 (No.898,520). Although a small illustrated advertisement for this cabinet had appeared in the April 1907 issue of the *Talking Machine World*, Mr. Solomon didn't apply for his patent until August 27, 1907. Manufactured by the Syracuse Cabinet Manufacturing Company, each drawer contains four stationary pegs, plus a circular "lazy Susan" with 57 pegs (305 total capacity). This example is marked "PAT AP'D FOR," suggesting it was manufactured prior to September 1908. Measures 28 1/4" square x 38" high. *Courtesy of Jalal and Charlotte Aro*. (Value code: VR)

4-80. This disc cabinet, with striking quarter-sawn oak veneer, contains both horizontal and vertical record storage. The square design was probably intended to support an external horn talking machine (internal horn instruments tended to be rectangular because of the necessity of encompassing the horn within the cabinet). The use of padded "slat" storage is reminiscent of Salter designs. The cabinet measures 22 1/2" wide, 22 1/2" deep and 34 1/2" high. *Courtesy of Fred Guenther.* (Value code: A)

4-81. This imposing oak double-chambered disc storage cabinet by the Salter Manufacturing Company employs the firm's characteristic horizontal dividers. As an advertisement illustrating this model (style No.784) in the November 1908 *Talking Machine World* proclaimed, "Records lay flat to keep them FROM WARPING." 29" wide, 18" deep and 38" high. *Courtesy of Walt and Ellen Bellm.* (Value code: A)

4-82. No less formidable is the cylinder version of the Salter cabinet in the previous illustration. Note that groups of scooped wooden posts on each shelf position the records in their original boxes, 198 total. As a November, 1908 advertisement explained, "Our new Cylinder Cabinets will hold ALL MAKES of Records in the original carton boxes." The ad illustrated Style "No.785," which was identical to the piece shown here, and accommodated Columbia Twentieth Century (6" long) cylinders on the bottom level. *Courtesy of Walt and Ellen Bellm.* (Value code: A)

4-83. The bulbous Salter Cabinet "No.786" presented a regal appearance, even without a talking machine perched on its top. When this model appeared in 1910, The Salter Manufacturing Company asserted that the popularity of its cabinets, "...which for the past two years have been the favorite among dealers on account of their superior quality to all makes, has kept our factory working overtime." See next illustration for open view. Measures 28" wide x 22 1/2" deep x 37 1/2" high. *Courtesy of the Sanfilippo collection.* (Value code: A)

4-84. The Salter "No. 786" in opened position. Note that the patented horizontal shelves on the door are more widely spaced than the rear sections. This was probably to decrease weight on the hinges and maintain the cabinet's stability when the loaded door was opened. *Courtesy of the Sanfilippo collection.*

4-85. For those with spacious homes: a disc record cabinet with storage compartments on four sides. The Salter Manufacturing Company offered this behemoth in 1911, illustrating it in the April issue of the *Talking Machine World*. Note the small drawer for needles or accessories. This deluxe mahogany cabinet (No."764"), measuring 26 1/4" (left face), 28 1/4" (right face) and 36 1/2" high, was large enough to accommodate 338 discs, and a wide variety of talking machines. *Courtesy of the Domenic DiBernardo collection.* (Value code: A)

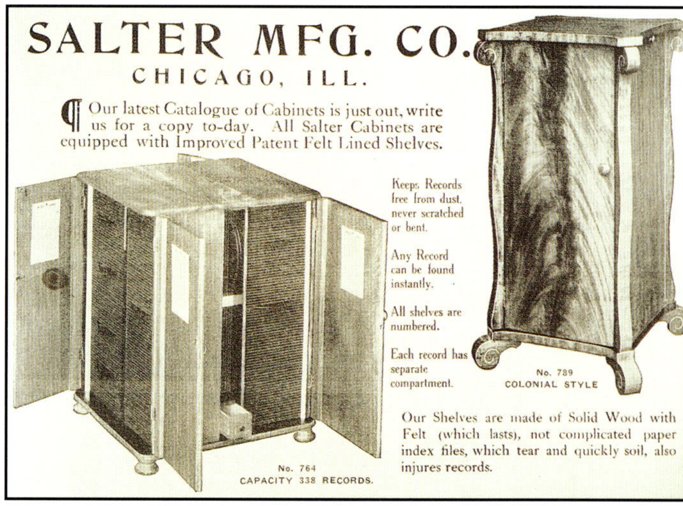

4-86. An advertisement from the April 1911 Talking Machine World featured the "No.764" cabinet.

4-87. The dimensions of this quarter-sawn oak record cabinet in the Empire style (which enjoyed a mild fad circa 1910; the Salter company referred to it as "Colonial") suggest it might hold either cylinders or discs – no doubt intentional, since it was produced at a time when these formats were competing strongly. 19 1/2" wide, 17 3/4" deep and 33" high. *Courtesy of Charles McCarn.* (Value code: B)

4-88. In this instance, simple shelves hold the kind of "scooped" wooden pegs seen in cabinets produced by the Salter Manufacturing Company. As shown here, cylinder boxes may be stored intact, or the records removed and positioned on the pegs. *Courtesy of Charles McCarn.*

4-89. This is a rather odd-looking record cabinet, with a restrained bit of routing on the door. It measures 14" wide, 14 1/2" deep and 41 1/2" high. *Courtesy of the collection of Howard Hazelcorn.* (Value code: VR)

4-90. The interior of the cabinet is no less unusual, and well thought-out. Scooped wooden pegs mounted on six revolving "caddies" hold eleven cylinder records each. Each circular level is interrupted by a gap to allow the records below it to be removed without obstruction. *Courtesy of the collection of Howard Hazelcorn.*

4-91. A very Herzog-looking cabinet, with gently bowed front and vivid oak grain, measuring 19 1/2" wide, 21" deep and 35 1/2" high. *Courtesy of the Sanfilippo collection.* (Value code: A)

4-92. Concealed behind the door are two disc record storage bins. In fact, Herzog manufactured "file drawer" type record cabinets under "Brown's patent." For an open view of Brown's design, see the next illustration. *Courtesy of the Sanfilippo collection.*

4-93. What appears to be an oak office filing cabinet is betrayed, in fact, by two key elements: the little drawer at the top right held steel needles, and the supplemental "foot" added support to the bottom drawer and kept the unit from pitching forward. Disc records were considerably heavier than office documents. The cabinet is marked, "Brown's Patent – May 3rd 1910, Herzog sole manufacturers for US." This was catalogue No. "D-45." 17" wide, 16 1/4" deep and 35" high. *Courtesy of the Sanfilippo collection.* (Value code: A)

4-94. Another "Brown's Patent" cabinet by Herzog, this one slightly larger – suitable for a Victor "V" talking machine, because of the shape of the applied pillars. *Courtesy of Jen and Ken Flaherty.* (Value code: A)

4-95. The interior, with Brown's filing system. Catalogue No "D-50." 21" wide, 19 1/2" deep and 36" high. *Courtesy of Jen and Ken Flaherty.*

4-96. The Herzog Art Furniture Company of Saginaw, Michigan, used the phrase "Record Filing Cabinet." In this instance, it really makes sense, since the style illustrated is based on the same premise as an office document filing cabinet. A Victor "III" fit perfectly on top. 3 1/2" x 6" folded.

Herzog Record Filing Cabinets

No. D 45

Golden Oak and Mahogany
Capacity—150 twelve-inch Discs.

Herzog Record Filing Cabinets

No. D 45, Open

Golden Oak and Mahogany.
Capacity—150 twelve-inch Discs.

It has remained for the HERZOG Cabinet to supply a record filing case, which at once provides safe protection for records and a natural support for the talking machine. HERZOG Cabinets fill every requirement accessory to the talking machine. Design, workmanship, construction and finish are conceded to be a triumph of ingenuity and skill and makes HERZOG Cabinets a desirable acquisition to any stock, when the best is none too good.

The HERZOG Disc Cabinets are provided with the celebrated Brown patent filing device, by which each record is filed separately between numbered cards and any record can be easily located by a ready reference record index book included with each cabinet. The heavy cards of the index file furnish a separate compartment for each record, to protect the sensitive sound grooves of the records from dust, and a patent spring prevents all rattling, scratching and warping.

Beautify your home with a HERZOG Cabinet, there is no home complete without one where a talking machine is owned.

4-97. Around 1910, "Flemish art" was all the rage. It was largely a hobby-based movement, with amateur craftspeople decorating wooden plaques, boxes and even furniture by pyrography, or wood burning. A prospective pyrographer could buy plans and tools from a craft supply house, and, hopefully, produce something more lasting than smoke. Flemish art employed outlined figures and designs against a mottled background that resembled a wood-burner's version of pointillism. The cylinder record cabinet shown here can be considered an excellent example of folk art. Measures 18 1/4" wide, 13 1/2" deep and 37 1/2" high, 15 cylinders on pegs per drawer. Interpreting the "blank" patch on the upper surface, a talking machine about the size of an Edison "Home" must have sat atop this cabinet. *Courtesy of Jim Ross.* (Value code: B)

4-98. The "paneled look" lasted only a short time in the Victor Talking Machine Company's catalogue. Groundbreaking table model Victrolas such as the second version of the "XII," the earliest versions of the "X" and "XI" (shown here) suggested the appearance of a cultured drawing room. Between 1910 and 1913 these models, with their juxtaposition of rounded molding and raised rectangular panels were textbook examples of early teens styling. Illustrated is a Victrola "XI" atop a record storage cabinet perfectly matched to its design. With flush molding, the two form a seamless unit Record cabinet measures 17 1/2" wide, 21 1/2" deep and 32" high. *Courtesy of Dale Lorang.* (Value code: B)

4-100. This interesting decal on the inner door surface of the cabinet in the previous illustration is similar to the commonly seen Victrola lid decals. *Courtesy of Fred Guenther*.

4-99. Victor offered record storage cabinets to fit its external and internal horn talking machines, but for some reason they seem to have been less popular than those sold by firms such as Herzog or Pooley. Shown here is a Victor cabinet from the early teens designed to fit a table model Victrola, 20 1/4" wide, 20 1/4" deep, 33 1/2" high. *Courtesy of Fred Guenther.* (Value code: A)

4-101. A Salter cabinet intended for the Victrola "XII" ("Baby Victrola," introduced 1909). Note how the major design elements of the Victrola – molding, corners, feet – are all tastefully reflected in the record cabinet. Measures 27" wide x 22 3/4" deep x 37" high. (Value code: A)

4-102. More than meets the eye: a simple table (measuring 20" wide x 24" deep x 32 3/4" high) provides support for a Victrola "XII" in uncatalogued quartered oak. Lyon & Healy was a prominent Chicago jobber of Victor products, and the "L & H Table" was announced in the July 1911 issue of the *Talking Machi World*. This table features a small retractable shelf just below the Victrola's sound doors, and record storage (in albums not shown) below One Lyon & Healy Disc Record Album was included with each table and more could be purchased for $1.50 each. The "L & H Table" sold to dealers for $7.00 and retailed at $10.00. The Victrola "XII" was introduced with a plain cabinet in July 1909 at a hefty $125.00. The more elaborate cabinet shown here was produced from January 1910 until the model's demise in September of the same year, and was not known to have been available in oak until this example (No. 4613A) was recently discovered. A Lyon & Healy decal on the lid suggests the possibil ity that this machine might have bee a special order. *Courtesy of Terry Paul Baer*. (Value code: C, table only)

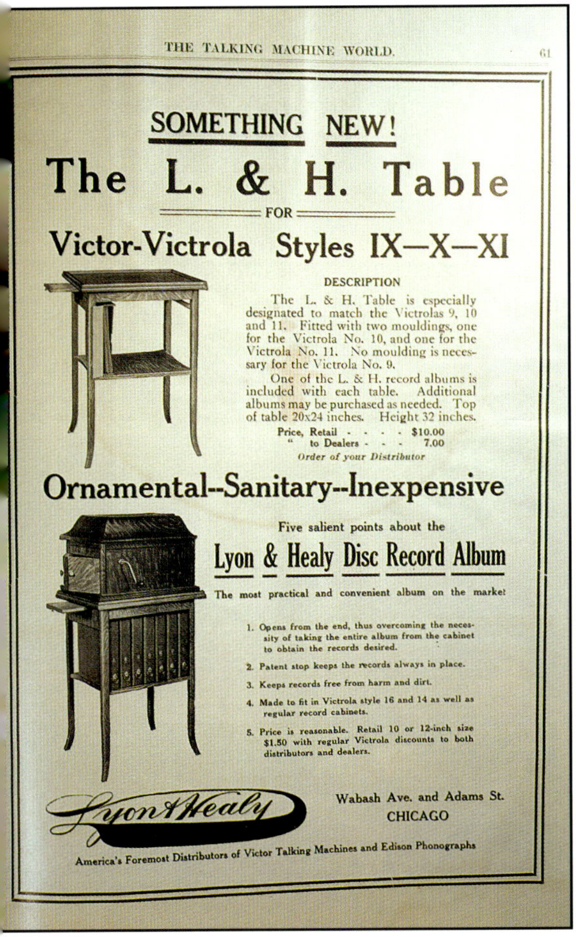
THE TALKING MACHINE WORLD. 61

SOMETHING NEW!

The L. & H. Table

FOR

Victor-Victrola Styles IX—X—XI

DESCRIPTION

The L. & H. Table is especially designated to match the Victrolas 9, 10 and 11. Fitted with two mouldings, one for the Victrola No. 10, and one for the Victrola No. 11. No moulding is necessary for the Victrola No. 9.

One of the L. & H. record albums is included with each table. Additional albums may be purchased as needed. Top of table 20x24 inches. Height 32 inches.

Price, Retail - - - - $10.00
" to Dealers - - - 7.00

Order of your Distributor

Ornamental--Sanitary--Inexpensive

Five salient points about the

Lyon & Healy Disc Record Album

The most practical and convenient album on the market

1. Opens from the end, thus overcoming the necessity of taking the entire album from the cabinet to obtain the records desired.
2. Patent stop keeps the records always in place.
3. Keeps records free from harm and dirt.
4. Made to fit in Victrola style 16 and 14 as well as regular record cabinets.
5. Price is reasonable. Retail 10 or 12-inch size $1.50 with regular Victrola discounts to both distributors and dealers.

Lyon & Healy

Wabash Ave. and Adams St.
CHICAGO

America's Foremost Distributors of Victor Talking Machines and Edison Phonographs

4-103. We suppose "sanitary" was not the first descriptor to pop into the heads of most record album buyers, but two out of three wasn't bad. This advertisement from the July 1911 *Talking Machine World* featured the oak table shown in the previous illustration, demonstrating how the shelf could be used for record storage.

4-104. A Columbia-marked disc record cabinet with an unusual variation of the "continuation top." A table model Grafonola, such as a "Favorite," was contained by the four corner moldings, giving the two pieces a complimentary appearance. 22 1/2" square x 36" high, holding 75 12" discs horizontally in the center, and 24 12" discs on the sides. (Value code: A)

4-105. The Keen-O-Phone was the brainchild of Morris Keen of Philadelphia. Beginning fitfully in 1910, the Keen-O-Phone brand began to develop. The first commercial production instrument was the "XXX" ("30"), which embodied the inventive vision of Morris Keen in a high-end ($210.00) cabinet model. The mechanism contained many unique technical features, however the disc record storage was especially notable. In this very early example of the "XXX," drawers similar to those employed in certain Edison Disc Phonograph models hold 78s in diagonal dividers. Keen-O-Phone cabinetry was executed by the Pooley Furniture Company of Philadelphia. *Courtesy of Brady Jeffcoat.*

4-106. The Pooley Furniture Company of Philadelphia manufactured a variety of record cabinets with its interesting record selection system. Usually, these cabinets hid the records behind wooden doors, but this model employed glass panels to display the selection mechanism and records. Measures 28 1/4" wide, 19 1/8" deep, and 45 1/2" high. *Courtesy of the Domenic DiBernardo collection.* (Value code: B)

4-107. A simple cabinet, holding 100 cylinder records, but delightful in appearance – from its cabriole legs to its graceful bail pulls to the decorative reliefs on its side panels. 15 3/4" wide, 18 3/4" deep and 27 1/2" high. (Value code: B)

4-108. We are seldom given the opportunity to step into the past. Dressing up in a top hat or visiting the local "Pioneer Village" give us amusing tastes. Gazing at an old picture of the house in which we presently live gets us closer to the real thing. In this photograph, an Edison "Triumph" Model "F" atop a stylish record cabinet dominates a 1912 interior. *Courtesy of Pat and Jamie Jones.*

4-109. Uncannily conjured up before our eyes, the very same "Triumph" becomes three-dimensional and warm with color. The cabinet measures 24 1/2" wide, 17 1/2" deep and 32 1/2" high. Each drawer holds 40 cylinders. *Courtesy of Pat and Jamie Jones.* (Value code: A, cabinet only)

4-110. The Vitaphone Company of Plainfield, New Jersey manufactured a line of disc talking machines in the early teens, based on an unconventional design premise. Vitaphone did not employ a soundbox that traveled across the record. Instead, the soundbox was stationary, and the vibrations were transferred to it via a solid length of "violin wood" that functioned like a tone arm. The system actually worked, but the company was under-capitalized and failed to make a go of it. Shown here, from the brief lifespan of the Vitaphone, are record albums for 12" diameter discs. Each album holds 15. *Courtesy of the Sanfilippo collection.* (Value code: E)

4-111. This unusual circular disc record cabinet holds 99 records. Inserting one's fingers in the horizontal slot just below the top rotates the internal carousel. The desired record is removed as shown. The indexed record number is visible through the opening at the top (see next illustration). Measures 29 3/4" diameter x 30" high. *Courtesy of Freddie and Jeanette Blair*. (Value code: B)

4-112. The top of the round cabinet shown in the previous illustration features a small viewer which displays the index number of the record ready to be accessed. *Courtesy of Freddie and Jeanette Blair.*

4-113. A modest but charming oak cabinet designed to fit an Edison Amberola "D-X" (introduced 1914). The routed designs suggest economy of manufacture, but a wealth of imagination. In the simplest configuration of the "continuation top," four recesses in the cabinet's upper surface accommodate the Phonograph's rubber bumpers. Clips on two modest shelves grip 36 Edison Blue Amberol record boxes. 13 5/8" wide, 17" deep and 30" high. *Courtesy of Walter and Carol Myers.* (Value code: B, cabinet only)

4-114. The same cabinet in the previous illustration, opened. *Courtesy of Walter and Carol Myers.*

4-115. Most disc records were sold in 10" or 12" diameter sizes. However, there were many variations that would not fit conveniently into conventional record storage. In 1906, Victor was offering 7", 8", 10" and 12" discs. In the teens, 8" and 9" discs would be revived. So, why not an *adjustable* storage cabinet? As seen here, it could be accomplished simply, by allowing the shelves to slide into a variety of positions. Measures 17" wide, 16 3/4" deep and 36 3/4" high. *Courtesy of the Scott and Denise Corbett collection.* (Value code: B)

4-116. Lundstrom of Little Falls, New York, sold "Converto" cabinets specially designed to enclosed table model Victrolas "IV" and "VI." By this means, the budget-range Victrolas, which were not supplied with dust-repelling lids, became all-inclusive units, complete with lids and record storage. Why, it was so easy, the lady of the house could do it! *Courtesy of Ron Seavers.*

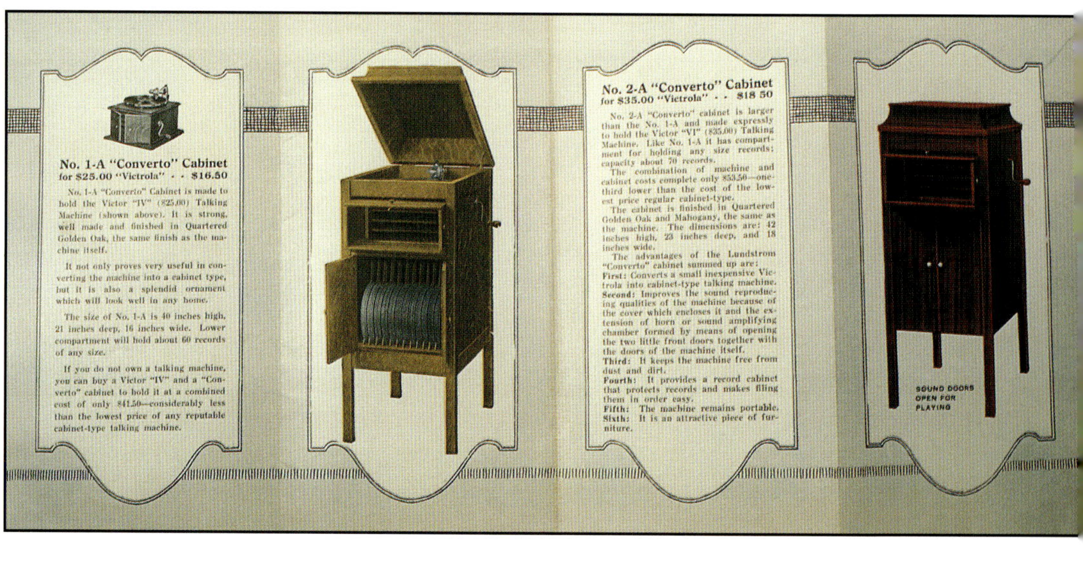

4-117. The horizontal upper doors of the "Converto" helped channel the sound from the Victrola nestled within. Lundstrom provided a popular option for those who initially could not afford the price of a full-sized talking machine. *Courtesy of Ron Seavers.*

4-118. Sectional or "stacking" bookcases, made up of modular units that could be built up as necessary, were very popular in the teens. Across the front of each unit there was usually a glass door that swung up and inward to expose the interior. The concept was naturally extended to disc record storage, as seen here. (A 1913 Lyon & Healy advertisement referred to them as "Unit Record Cabinets.") This fumed oak cabinet, measuring 34 3/4" wide, 13 1/2" deep and 44 1/2" high, has a definite Arts and Crafts look. It is made up of two stacking units, the top one for 10" discs, the bottom for 12", 90 slots per tier. The slots are wide enough to accept Edison Diamond Disc records (which were approximately twice as thick as conventional 78s). *Courtesy of Charles McCarn.* (Value code: B)

4-119. This one-piece mahogany disc record storage cabinet mimics the "sectional" look, but with side-hinged doors. The chamfered molding and leaded glass elegantly mask conventional vertical record storage (the lower two tiers for 12" diameter discs). 36 1/2" wide, 16" deep and 57 1/4" high, holding 277 records. *Courtesy of Charles McCarn.* (Value code: VR)

4-120. The Wood Manufacturing Company of Los Angeles produced a high quality line of disc record cabinets. Some were designed with recessed tops (shown) to blend with the molding of a talking machine (in this instance a Victrola from the early teens). Features of the cabinet include cloth covered wood dividers to separate the discs, and a push-button record retrieval system different from the one employed by Pooley. The push button-concept occurred to others, too – in 1907 the Automatic Disc Holder Company of New York City advertised apparatus for installation in disc record cabinets, "You have but to *press the key* and the mechanism rolls it [the record] slowly out into your hand." Since internal horn talking machines usually were deeper than they were wide, most disc storage manufacturers sacrificed the rear portion of the cabinet to dead, unusable space. Here, however, a unique side door (resembling something out of the imagination of Lewis Carroll) accesses an additional cache of records. The small drawer at the top right held the index (shown above), accessories and needles. *Courtesy of Fred Guenther.* (Value code: A)

4-121. This decal from the interior of one of the doors identified the "Wood Filing Cabinet" line of record storage cabinets. *Courtesy of Fred Guenther.*

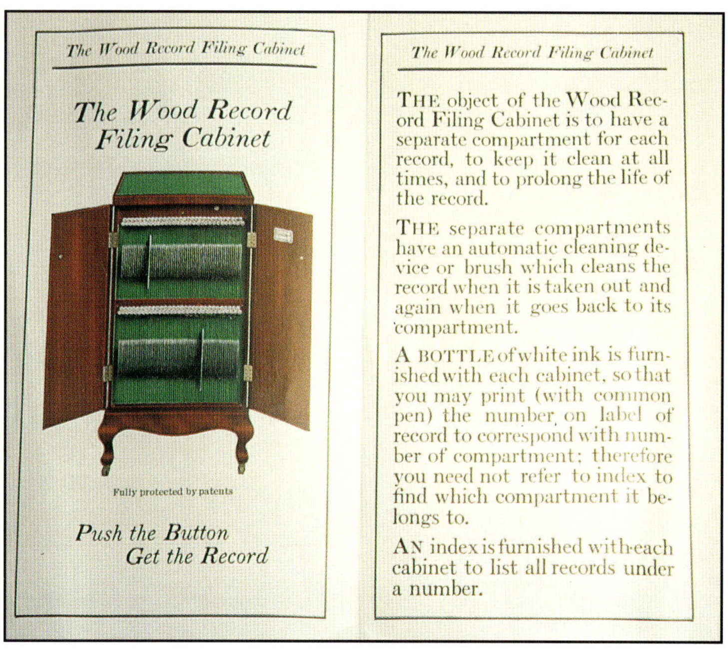

The Wood Record Filing Cabinet

The Wood Record Filing Cabinet

Fully protected by patents

Push the Button Get the Record

The Wood Record Filing Cabinet

THE object of the Wood Record Filing Cabinet is to have a separate compartment for each record, to keep it clean at all times, and to prolong the life of the record.

THE separate compartments have an automatic cleaning device or brush which cleans the record when it is taken out and again when it goes back to its compartment.

A BOTTLE of white ink is furnished with each cabinet, so that you may print (with common pen) the number on label of record to correspond with number of compartment; therefore you need not refer to index to find which compartment it belongs to.

AN index is furnished with each cabinet to list all records under a number.

4-122. A flyer for the "Wood Record Filing Cabinet" features the well-conceived slogan "Push the Button, Get the Record." Note the instructions to index the label of each record with a filing number in white ink. Modern record collectors would dream of finding the guy responsible for promoting the desecration of record labels, and meting out to him summary justice! *Courtesy of the Scott and Denise Corbett collection.*

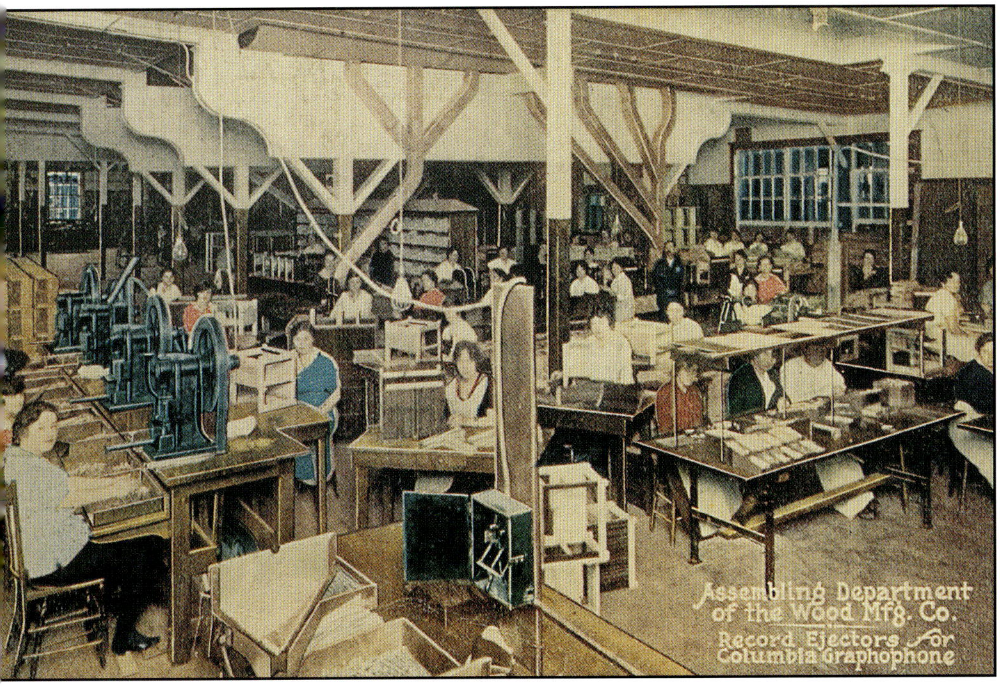

4-123. The factory interior of the Wood Manufacturing Company of Los Angeles, circa 1920. Note the reference to Columbia Graphophones. Grafonolas (the coined name for Columbia internal horn Disc Graphophones) are sometimes found equipped with Wood Manufacturing's record ejectors. *Courtesy of Doug Anderson.*

4-124. This rather plain cylinder record cabinet is just the right size to support an Edison Amberola "30." It disguises well its interesting secret. Records are stored in sliding trays. *Courtesy of Philippe Le Ray.* (Value code: C)

4-125. This is the same cabinet, and the secret is out – it comes apart completely, with no effort. Was the purpose to make it easy to transport? Just as likely, it was supposed to be easy to assemble. This cabinet was probably sold as a "kit" for the enterprising homeowner. *Courtesy of Philippe Le Ray.*

4-126. In the quest for portability, many elements of a talking machine could be minimized, but the universality of the 10" diameter disc continued to impose physical limits on design. The shape of this cleverly constructed example of individual craftsmanship is defined by the records it houses. *Courtesy of John W. Hess.* (Value code: D)

4-127. In this custom-built portable talking machine, all features have been pared down except the conventional mechanical parts, circa 1920. Two "wings" on which new and used needle cups have been mounted pivot outward, also revealing the mouth of a rudimentary internal horn. The winding crank (not shown) stores in one of the side baskets. *Courtesy of John W. Hess.*

4-128. This cabinet, manufactured by the Long Company of Hanover, Pennsylvania, would have been a good match for a Victrola "IX" or a similar machine. The door doubled as a sturdy shelf on which records might be sorted. Measures 24" wide x 20" deep x 30" high. *Courtesy of Freddie and Jeanette Blair.* (Value code: C)

4-129. The marketing phrase "continuation top" had been used early-on to denote a cabinet whose lines were meant to flow into those of the machine for which it was intended. The later version of the popular Victrola "IX" presented a challenge with its curvaceous apron. This unmarked cabinet is evidence that at least one manufacturer was equal to the task. Measures 19 1/2" wide, 23" deep, and 33" high. *Courtesy of the Sanfilippo collection*. (Value code: B, cabinet only)

4-130. It's unusual to find a bombé base cabinet designed to fit a moderately-priced Victrola such as the VV "IX," circa 1920. Note the corner indentations that accepted the feet of the instrument. The first version of the "IX" (introduced 1911) was not elevated on feet. This cabinet was likely the work of the Salter Manufacturing Company of Chicago, which made horizontal record storage of the type shown here its signature styling. An advertisement from the September 1910 *Talking Machine World* explained, "The interiors of Salter Disc Cabinets have an individual compartment for each record. The wood shelves are lined with soft green felt to prevent records from scratching." Measures 21" wide, 17" deep and 32" high. *Courtesy of Richard and Nancyann Brown.* (Value code: B)

4-131. As the internal horn talking machine evolved, its cabinet took on more importance. During the late teens and early 1920s, all the major manufacturers offered "art cases" and custom cabinetry for those who could afford such extravagances. This immense (53 1/2" wide, 15 1/2" deep and 42 1/2" high) cabinet, holding 600 discs, could have complimented one of the expensive behemoths produced by Victor, Columbia. Edison or Brunswick during this period. There are three vertical compartments, the topmost exclusively for 10" diameter records. *Courtesy of Gilbert "Bert" Pasley."* (Value code: A)

Certain underlying elements are present among all the accessories and contraptions that we have illustrated: tangible evidence of intelligence, a yearning for beauty, a need for order. All this can be discovered in phonograph collections around the world and within the pages of this book. The human element has been emphasized by us before, and rather than belaboring the point, we would encourage the reader to turn back to a random page, pick an item, and let it speak to you. We would furthermore endorse a hands-on approach. Go to an antique show. Pick up a gizmo that catches your eye. Or simply get out your favorite record and play it right now.

HELPFUL INFORMATION

Atmospheric changes and conditions encountered by furniture while in transit and, often, in the home cause new wood to swell, so that drawers and doors bind and stick.

In winter the dry heat of the house will generally ease the working parts within a few weeks and it is better to avoid dressing down the wood, because it is designed to fit perfectly when dry. If in the summer it seems too long to wait for a natural shrinkage, which will occur when the house is heated again, anyone with a good plane may, if careful, dress the parts that bind just enough to make action free. But never use the plane on drawers until you have withdrawn them from the case, turned them up and examined lower-edge of front, also the tops of stationary rails between drawers, for possible excess of varnish which should be scraped off with knife or glass. Before planing a door-edge take a screwdriver and make sure the hinge-screws were driven quite in so that hinges close tightly; examine all four edges of door, particularly the hinged-edge for excess-varnish. Often a little Sweet Home Soap applied to the bearing points will make the drawers work all right.

EVERY new piece of polished furniture should be gone over with Larkin Furniture polish to remove any chafing, scratches and the dust of travel; waxed-finished furniture should receive an application of Larkin Floor Wax.

Drawer pulls, casters and all trimmings, if any, will be found, packaged, in a drawer, if any. Key will be found attached to outside of case.

To open a drawer that sticks. Place against the corner-post of case at end of drawer a small block of wood over a heavy pad of paper. Strike block a sharp blow with a hammer. Repeat this at each end alternately until drawer will open.

THIS ARTICLE MANUFACTURED BY OUR FACTORY No. 15

If you have occasion to write us about it, mention this number without fail. Do not cause useless delay by omitting this number.

Larkin Co.
BUFFALO, N. Y.

4-132. This informative label is affixed to the underside of a disc record cabinet sold by the Larkin Company of Buffalo, New York. During the 1920s, Larkin manufactured or distributed a variety of products for the home, including furniture and talking machines. The text is as relevant today as it was 80 years ago – what modern collector has not encountered swollen doors or balky drawers? *Courtesy of Charles McCarn.*

BRUNSWICK *Topics*

Now What Songs Could They Be?

Eight song titles are illustrated here---try to guess what they are---it's lots of fun. And if you grow tired of trying (you really shouldn't, cause they're easy), just look over the lists throughout this issue and you'll find them all there. Easy, eh!

4-133. From the "house organ" *Brunswick Topics*, December 12, 1928 – how good is YOUR knowledge of vintage American pop music? *Courtesy of Martin F. Bryan.*

Bibliography

Association for Recorded Sound Collections, various eds. *ARSC Journal*, various issues to present.

Barnum, Frederick O., III. *His Master's Voice in America*, Camden, General Electric Company, 1991.

Barr, Steven C.: *The Almost Complete 78 RPM Record Dating Guide (II)* Huntington Beach: Yesterday Once Again, 1992.

Baumbach, Robert W. *Columbia Phonograph Companion, Vol. II (Disc Graphophones and the Grafonola)*. Woodland Hills: Stationery X-Press, 1996.

Baumbach, Robert W. *Look for the Dog*. Woodland Hills: Stationery X-Press, 1996.

Bayly, E., ed. *The EMI Collection*. Bournemouth (GB): Talking Machine Review, 1977.

_____. *The Talking Machine Review International*. Bournemouth (GB): various issues beginning 1970.

Bayly, Ernie and Michael Kinnear. *The Zon-o-phone Record*. Heidelberg (Victoria, Australia): Michael Kinnear, 2001.

Bill, Edward Lyman, ed. *The Talking Machine World*. New York: various issues 1905-1929.

Brooks, Tim, ed. *Little Wonder Records, a History and Discography*. St. Johnsbury (VT): New Amberola Phonograph Company, 2000.

Bryan, Martin, ed. *The New Amberola Graphic*. St. Johnsbury (VT): various issues 1973 to present.

Charosh, Paul, compiled by. *Berliner Gramophone Records, American Issues 1892-1900*. Westport: Greenwood Press, 1995.

City of London Phonograph and Gramophone Society, various eds. *The Hillandale News* (beginning in 2002 *For the Record*). various issues 1960s to present.

Conot, Robert. *A Streak of Luck*. New York: Seaview Books, 1979.

Dethlefson, Ronald, ed. *Edison Blue Amberol Recordings, 1912-1914*. Brooklyn: APM Press, 1980.

_____. *Edison Blue Amberol Recordings, 1912-1914* (paper cover). Woodland Hills (CA): Stationery X-Press, 1997.

_____. *Edison Blue Amberol Recordings, 1915-1929*. Brooklyn: APM Press, 1981.

_____. *Edison Blue Amberol Recordings, 1915-1929*, second edition. Los Angeles: Mulholland Press, 1999.

Dethlefson, Ronald and George Copeland. *Pathé Records and Phonographs in America, 1914-1922*. Los Angeles: Mulholland Press, 2000.

Dixon, Robert and John Godrich. *Recording the Blues*. New York: Stein and Day Publishers, 1970.

Edge, Ruth and Leonard Petts. *The Collector's Guide to 'His Master's Voice' Nipper Souvenirs*. London: EMI Group, 1997.

Fabrizio, Timothy C. and George F. Paul. *Antique Phonograph Advertising: An Illustrated History*. Atglen (PA): Schiffer Publishing, Ltd., 1999.

_____. *Antique Phonograph Gadgets, Gizmos and Gimmicks*. Atglen (PA): Schiffer Publishing, Ltd., 1999.

_____. *Discovering Antique Phonographs*. Atglen (PA): Schiffer Publishing, Ltd., 2000.

_____. *Phonographs With Flair, a Century of Style In Sound Reproduction*. Atglen (PA): Schiffer Publishing, Ltd., 2001.

_____. *The Talking Machine, an Illustrated Compendium 1877-1929*. Atglen (PA): Schiffer Publishing, Ltd., 1997.

Fagan, Ted, compiled by. *The Encyclopedic Discography of Victor Recordings, Pre-Matrix Series*. Westport: Greenwood Press, 1983.

_____. *The Encyclopedic Discography of Victor Recordings*. Westport: Greenwood Press, 1986.

Frow, George L. *Edison Cylinder Phonograph Companion*. Woodland Hills: Stationery X-Press, 1994.

_____. *The Edison Disc Phonographs and the Diamond Discs*. Sevenoaks, Kent (GB): George L. Frow, 1982.

_____. *The Edison Disc Phonographs and the Diamond Discs*. Los Angeles: Mulholland Press, paperback reprint 2001.

Gaisberg, Frederick W. *The Music Goes Round*. North Stratford (NH): Ayer Company Publishers, Inc., 1977 (1942 reprint).

Gelatt, Roland. *The Fabulous Phonograph, revised*. New York: Appleton-Century, 1965 (three eds. 1955, 1977).

Gracyk, Tim, with Frank Hoffmann. *Popular American Recording Pioneers, 1895-1925*. Binghamton (NY): Haworth Press, 2000.

Gracyk, Tim, ed. *Victrola and 78 Journal*. Roseville (CA): various issues 1994-1998.

Hatcher, Danny, ed. *Proceedings of the 1890 Convention of Local Phonograph Companies*. Nashville: Country Music Foundation Press, reprint, 1974.

Hazelcorn, Howard. *A Collector's Guide to the Columbia Spring-Wound Cylinder Graphophone, 1894-1910*. Brooklyn: APM Press, 1976.

_____. *Columbia Phonograph Companion, Volume I: Hazelcorn's*

Guide to the Columbia Cylinder Graphophone. Los Angeles: Mulholland Prcss, 1999.

Hunting, Russell, ed. *The Phonoscope.* New York: Phonoscope Publishing Company, various issues 1896-1900.

Johnson, E. R. Fenimore. *His Master's Voice Was Eldridge R. Johnson.* Milford (DE): State Media, Inc., 1975.

Koenigsberg, Allen, ed. *The Antique Phonograph Monthly.* Brooklyn: APM Press, various issues 1972-1993.

_____. *Edison Cylinder Records, 1889-1912, 2nd ed.* Brooklyn: APM Press, 1988.

_____. *The Patent History of the Phonograph, 1877-1912.* Brooklyn: APM Press, 1991.

Laird, Ross. *Brunswick Records, a Discography of Recordings 1916-1931*, in four volumes. Westport: Greenwood Publishing Group, 2001.

Lorenz, Kenneth M. *Two-Minute... Cylinders of the Columbia Phonograph Company.* Wilmington (DE): Kastlemusick, Inc., 1981.

Maken, Neil. *Hand-Cranked Phonographs.* Huntington Beach: Promar Publishing, 1995.

Marco, Guy A. and Frank Andrews, eds. *Encyclopedia of Recorded Sound in the United States.* New York: Garland Publishing, Inc., 1993.

Martland, Peter: *Since Records Began: EMI: the First Hundred Years.* (Distributed by Timber Press, Portland, OR): Amadeus Press, 1997.

Marty, Daniel. *Histoire Illustrée du Phonographe (Illustrated History of the Phonograph).* Lausanne/Paris: Edita/Lazarus, 1979, (reprinted in various English language editions with slight modifications of the title).

Michigan Antique Phonograph Society, various eds. *In the Groove.* Various issues to present.

Moore, Jerrold Northrop. *A Matter of Records.* New York: Taplinger Publishing Company, 1977.

Moore, Wendell, ed. *The Edison Phonograph Monthly*, various anthologies 1903-1916. New Albany (IN): Wendell Moore Publications.

Proudfoot, Christopher. *Collecting Phonographs and Gramophones.* Ncw York: Mayflower Books, 1980.

Read, Oliver and Walter L. Welch. *From Tinfoil to Stereo.* Indianapolis: Howard W. Sams and Company, Inc., 1959.

Reiss, Eric L. *The Compleat Talking Machine.* Chandler (AZ): Sonoran Publishing, (revised)1998.

Rust, Brian, compiled by. *Discography of Historical Records on Cylinders and 78s.* Westport: Greenwood Press, 1979.

Rust, Brian and Tim Brooks: *Columbia Master Book Discography*, in four volumes, Westport, Greenwood Publishing Group, 1999.

Seiter, Bessie & Floyd, and Robin & Joan Rolfs. *Phonograph Dolls That Talk and Sing.* Hortonville (WI): Audio Antique LLC, 2001.

Sherman, Michael W. (with William R. Moran and Kurt R. Nauck, III). *Collector's Guide to Victor Records.* Dallas: Monarch Record Enterprises, 1992.

Sherman, Michael W. and Kurt R. Nauck, III. *Note the Notes, An Illustrated History of the Columbia 78 rpm Record Label, 1901-1958.* New Orleans: Monarch Record Enterprises, 1998.

Stewart-Baxter, Derrick. *Ma Rainey and the Classic Blues Singers.* New York: Stein and Day Publishers, 1970.

Sutton, Allan. *Directory of American Disc Record Brands and Manufacturers, 1891-1943.* Westport: Greenwood Press, 1994.

Sutton, Allan and Kurt Nauck. *American Record Labels and Companies, an Encyclopedia (1891-1943)* [with CD-ROM]. Denver: Mainspring Press, 2000.

Welch, Walter L., and Leah Brodbeck Stenzel Burt. *From Tinfoil to Stereo, 1877-1929.* Gainesville: University Press of Florida, 1994.

Wile, Raymond R.and Ronald Dethlefson. *Edison Diamond Disc Recreations, Records and Artists 1910-1929.* Brooklyn: APM Press, 1985.

Glossary

AMBEROL: The first four-minute cylinder developed by Edison, 1908-1912. These cylinders are black in color, made of especially hard metallic soap, and were usually sold in green and white containers.

AMBEROLA: The name denoting Edison's line of internal-horn cylinder Phonographs, 1909-1929.

BACK MOUNT: A term used to describe a talking machine which uses a back bracket which supports the horn and thus removes its inertia from the SOUNDBOX.

BEDPLATE: The metal motor mounting plate (often a casting) to which the upper mechanism of a cylinder talking machine is mounted, and from which the motor is suspended.

BLANK: A smooth, grooveless unrecorded cylinder.

BLUE AMBEROL: The blue celluloid four-minute cylinder sold by Edison, 1912-1929. The containers for these cylinders were blue until 1917, orange and blue thereafter.

CARRIAGE: The assembly which "carries" the REPRODUCER of a cylinder talking machine across the recording.

CARRIER ARM: Edison nomenclature for the CARRIAGE. The HALF-NUT and its flat supporting spring are attached directly to the carrier arm on nearly all Edison machines.

COMBINATION ATTACHMENT: Devices offered by Edison to convert pre-1908 Phonographs to play four-minute cylinders in addition to the two-minute variety.

CONCERT: Edison's trade name for the five-inch diameter cylinder and the Edison Phonograph designed to play it.

CRANE: The support used to mount horns of 15-inch length or greater to cylinder talking machines.

CYLINDER: A geometric form on which entertainment recordings were made, 1877-1929. The earliest sheets of tinfoil were followed by self-supporting cylinders of ozocerite-covered cardboard, stearic acid/paraffin, hard metallic soap, and celluloid. Advantages of the cylinder format included a constant surface speed and the ability to make home recordings.

DIAMOND DISC: Edison's line of disc records and machines, 1912-1929. The Diamond Discs were vertically recorded, and the Diamond Disc Phonographs were one of the very few disc talking machines to use a FEEDSCREW.

DIAPHRAGM: The flexible, circular vibrating membrane of a SOUNDBOX or REPRODUCER which converts mechanical energy to acoustic energy or sound waves. Mica, glass, compressed paper, copper and aluminum were commonly used as diaphragms in early talking machines.

ELBOW: The connection between the horn and the

SOUNDBOX or TONE ARM of a disc talking machine. Earliest elbows are made of leather, gradually giving way to metal elbows after 1900.

ENDGATE: A swinging arm which supports one end of the MANDREL on some cylinder talking machines. To mount or remove a cylinder it is therefore necessary to open and close the endgate.

FEEDSCREW: A threaded rod which usually drives a half-nut fixed to the carriage of a cylinder talking machine, thus guiding the reproducer across the record grooves. Certain disc machines used FEEDSCREWS to drive the SOUNDBOX across the record or to move the turntable beneath a fixed soundbox. Similarly, some cylinder machines used feedscrews to drive a mandrel longitudinally beneath a fixed reproducer.

FRONT MOUNT: A term used to describe a disc machine where the horn attaches directly to the SOUNDBOX, and the support arm (or mount) runs directly below and parallel to the HORN. In such an arrangement the support arm points in the same general direction as the bell of the horn.

GRAMOPHONE: Originally the name by which Emile Berliner's disc talking machine was known, it became a generic term to denote any disc playing talking machine, but fell out of use in the United States.

GRAND: Columbia's trade name for the five-inch diameter cylinder and the various Columbia machines designed to play it.

GOVERNOR: The mechanical assembly in a talking machine motor which regulates the speed, usually by limiting the outward movement of spinning weights.

HALF-NUT: An internally threaded metal piece usually in the form of a nut which has been cut in half. The threads of the half-nut correspond to the threads of the FEEDSCREW. As the feedscrew revolves, the half-nut will be propelled along it, thus driving the CARRIAGE of a talking machine.

LATERAL RECORDING: The type of disc talking machine recording where the information is encoded in sides of the groove. (see VERTICAL RECORDING)

MANDREL: The tapered drum upon which the cylinder record is placed for playing.

NEEDLE: The point (usually steel) of the SOUNDBOX which rides the grooves of the recording and transmits vibrations to the DIAPHRAGM.

NEEDLE HEAD: A colloquial term for the SOUNDBOX or PICK-UP which accepts vibrations directly from the record.

NEEDLE TIN: A container to store and dispense steel talking machine needles.

NEEDLE SHARPENER (OR CUTTER): A device for repointing disc talking machine needles of fiber, thorn or steel.

PHONOGRAPH: Originally, the name given by Edison to the sound recording/reproducing device he invented in 1877. In American usage, the term has been applied to sound reproducing equipment in general, regardless of age or type of record played.

RECORDER: A device comprising a DIAPHRAGM and cutter stylus for recording cylinder or disc records.

REPEATING ATTACHMENT: A device for cylinder or disc talking machines which returns the NEEDLE or STYLUS to the beginning of the record after the selection has played.

REPRODUCER: The component comprising STYLUS, linkage and DIAPHRAGM, which reproduces the sound from the record grooves. This term is most frequently applied to cylinder machines. (See SOUNDBOX)

SHAVER: On a cylinder talking machine, a device which holds a knife in position to pare the recorded surface from a wax cylinder.

SOUNDBOX: The component comprising NEEDLE, linkage and DIAPHRAGM, which reproduces the sound from the record grooves. This term is most frequently applied to disc machines. (See REPRODUCER)

STYLUS: The point (usually sapphire or diamond) of the REPRODUCER which rides the grooves of the recording and transmits vibrations to the DIAPHRAGM or PICK-UP.

TALKING MACHINE: A general term for sound reproducing devices playing either cylinder or disc records, made prior to the advent of electric reproduction. Specific brands such as Graphophone, Victrola or Amberola all fall within the general category of talking machines. Modern American usage substitutes PHONOGRAPH for the older term "talking machine." For the purposes of historical accuracy, both terms will appear frequently in this book.

TONE ARM: A movable hollow tube which conducts sound to the HORN from the SOUNDBOX.

TRUNNION: Rightly, the sleeve visible on either side of the CARRIAGE which slides over the FEEDSCREW shaft on a Graphophone, but commonly used to denote the entire carriage.

TURNTABLE: The round platter which supports a disc record and revolves while the record is being played. After 1960, that separate component in a SOUND SYSTEM which comprised the TURNTABLE and RECORD CHANGER was often referred to simply as the TURNTABLE.

VERTICAL RECORDING: The type of cylinder or disc recording in which the sound vibrations are encoded in the bottom of the groove. (see LATERAL RECORDING)

WAX: Although the material from which many cylinder records were made is generally referred to as "wax," it was more rightly various formulations of metallic soap. The resemblance of this material to conventional wax promotes the common descriptive term.

Index